# OFFICIAL SQA SPECIMEN QUESTION PAPER AND HODDER GIBSON MODEL QUESTION PAPERS WITH ANSWERS

*get help with page 77 (Q 14b) from Jack*

## NATIONAL 5

# CHEMISTRY

## 2013 Specimen Question Paper & 2013 Model Papers

D0259513

This book contains the official 2013 SQA Specimen Question Paper for National 5 Chemistry, with associated SQA approved answers modified from the official marking instructions that accompany the paper.

In addition the book contains model practice papers, together with answers, plus study skills advice. These papers, some of which may include a limited number of previously published SQA questions, have been specially commissioned by Hodder Gibson, and have been written by experienced senior teachers and examiners in line with the new National 5 syllabus and assessment outlines, Spring 2013. This is not SQA material but has been devised to provide further practice for National 5 examinations in 2014 and beyond.

Hodder Gibson is grateful to the copyright holders, as credited on the final page of the Answer Section, for permission to use their material. Every effort has been made to trace the copyright holders and to obtain their permission for the use of copyright material. Hodder Gibson will be happy to receive information allowing us to rectify any error or omission in future editions.

Hachette UK's policy is to use papers that are natural, renewable and recyclable products and made from wood grown in sustainable forests. The logging and manufacturing processes are expected to conform to the environmental regulations of the country of origin.

Orders: please contact Bookpoint Ltd, 130 Park Drive, Abingdon, Oxon OX14 4SE. Telephone: (44) 01235 827720. Fax: (44) 01235 400454. Lines are open 9.00–5.00, Monday to Saturday, with a 24-hour message answering service. Visit our website at www.hoddereducation.co.uk. Hodder Gibson can be contacted direct on: Tel: 0141 848 1609; Fax: 0141 889 6315; email: hoddergibson@hodder.co.uk

This collection first published in 2013 by
Hodder Gibson, an imprint of Hodder Education,
An Hachette UK Company
2a Christie Street
Paisley PA1 1NB

BrightRED    Hodder Gibson is grateful to Bright Red Publishing Ltd for collaborative work in preparation of this book and all SQA Past Paper and National 5 Model Paper titles 2013.

Typeset by PDQ Digital Media Solutions Ltd, Bungay, Suffolk NR35 1BY

Printed in the UK

A catalogue record for this title is available from the British Library

ISBN: 978-1-4718-0199-0

3 2 1

2014 2013

# Introduction

## Study Skills – what you need to know to pass exams!

### Pause for thought

Many students might skip quickly through a page like this. After all, we all know how to revise. Do you really though?

### Think about this:

"IF YOU ALWAYS DO WHAT YOU ALWAYS DO, YOU WILL ALWAYS GET WHAT YOU HAVE ALWAYS GOT."

Do you like the grades you get? Do you want to do better? If you get full marks in your assessment, then that's great! Change nothing! This section is just to help you get that little bit better than you already are.

There are two main parts to the advice on offer here. The first part highlights fairly obvious things but which are also very important. The second part makes suggestions about revision that you might not have thought about but which WILL help you.

## Part 1

DOH! It's so obvious but …

### Start revising in good time

Don't leave it until the last minute – this will make you panic.

Make a revision timetable that sets out work time AND play time.

### Sleep and eat!

Obvious really, and very helpful. Avoid arguments or stressful things too – even games that wind you up. You need to be fit, awake and focused!

### Know your place!

Make sure you know exactly **WHEN and WHERE** your exams are.

### Know your enemy!

**Make sure you know what to expect in the exam.**

How is the paper structured?

How much time is there for each question?

What types of question are involved?

Which topics seem to come up time and time again?

Which topics are your strongest and which are your weakest?

Are all topics compulsory or are there choices?

### Learn by DOING!

There is no substitute for past papers and practice papers – they are simply essential! Tackling this collection of papers and answers is exactly the right thing to be doing as your exams approach.

## Part 2

People learn in different ways. Some like low light, some bright. Some like early morning, some like evening / night. Some prefer warm, some prefer cold. But everyone uses their BRAIN and the brain works when it is active. Passive learning – sitting gazing at notes – is the most INEFFICIENT way to learn anything. Below you will find tips and ideas for making your revision more effective and maybe even more enjoyable. What follows gets your brain active, and active learning works!

### Activity 1 – Stop and review

#### Step 1

When you have done no more than 5 minutes of revision reading STOP!

#### Step 2

Write a heading in your own words which sums up the topic you have been revising.

#### Step 3

Write a summary of what you have revised in no more than two sentences. Don't fool yourself by saying, 'I know it but I cannot put it into words'. That just means you don't know it well enough. If you cannot write your summary, revise that section again, knowing that you must write a summary at the end of it. Many of you will have notebooks full of blue/black ink writing. Many of the pages will not be especially attractive or memorable so try to liven them up a bit with colour as you are reviewing and rewriting. **This is a great memory aid, and memory is the most important thing.**

## Activity 2 — Use technology!

Why should everything be written down? Have you thought about 'mental' maps, diagrams, cartoons and colour to help you learn? And rather than write down notes, why not record your revision material?

What about having a text message revision session with friends? Keep in touch with them to find out how and what they are revising and share ideas and questions.

Why not make a video diary where you tell the camera what you are doing, what you think you have learned and what you still have to do? No one has to see or hear it but the process of having to organise your thoughts in a formal way to explain something is a very important learning practice.

Be sure to make use of electronic files. You could begin to summarise your class notes. Your typing might be slow but it will get faster and the typed notes will be easier to read than the scribbles in your class notes. Try to add different fonts and colours to make your work stand out. You can easily Google relevant pictures, cartoons and diagrams which you can copy and paste to make your work more attractive and **MEMORABLE**.

## Activity 3 – This is it. Do this and you will know lots!

### Step 1

In this task you must be very honest with yourself! Find the SQA syllabus for your subject (www.sqa.org.uk). Look at how it is broken down into main topics called MANDATORY knowledge. That means stuff you MUST know.

### Step 2

BEFORE you do ANY revision on this topic, write a list of everything that you already know about the subject. It might be quite a long list but you only need to write it once. It shows you all the information that is already in your long-term memory so you know what parts you do not need to revise!

### Step 3

Pick a chapter or section from your book or revision notes. Choose a fairly large section or a whole chapter to get the most out of this activity.

With a buddy, use Skype, Facetime, Twitter or any other communication you have, to play the game "If this is the answer, what is the question?". For example, if you are revising Geography and the answer you provide is "meander", your buddy would have to make up a question like "What is the word that describes a feature of a river where it flows slowly and bends often from side to side?".

Make up 10 "answers" based on the content of the chapter or section you are using. Give this to your buddy to solve while you solve theirs.

### Step 4

Construct a wordsearch of at least 10 X 10 squares. You can make it as big as you like but keep it realistic. Work together with a group of friends. Many apps allow you to make wordsearch puzzles online. The words and phrases can go in any direction and phrases can be split. Your puzzle must only contain facts linked to the topic you are revising. Your task is to find 10 bits of information to hide in your puzzle but you must not repeat information that you used in Step 3. DO NOT show where the words are. Fill up empty squares with random letters. Remember to keep a note of where your answers are hidden but do not show your friends. When you have a complete puzzle, exchange it with a friend to solve each other's puzzle.

### Step 5

Now make up 10 questions (not "answers" this time) based on the same chapter used in the previous two tasks. Again, you must find NEW information that you have not yet used. Now it's getting hard to find that new information! Again, give your questions to a friend to answer.

### Step 6

As you have been doing the puzzles, your brain has been actively searching for new information. Now write a NEW LIST that contains only the new information you have discovered when doing the puzzles. Your new list is the one to look at repeatedly for short bursts over the next few days. Try to remember more and more of it without looking at it. After a few days, you should be able to add words from your second list to your first list as you increase the information in your long-term memory.

## FINALLY! Be inspired...

Make a list of different revision ideas and beside each one write **THINGS I HAVE** tried, **THINGS I WILL** try and **THINGS I MIGHT** try. Don't be scared of trying something new.

And remember – "FAIL TO PREPARE AND PREPARE TO FAIL!"

# National 5 Chemistry

## The course

Before sitting your National 5 Chemistry examination, you must have passed three **Unit Assessments** within your school or college, and produced an additional short report (approximately 100 words).

To achieve a pass in National 5 Chemistry there are then two further main components.

### Component 1 – Assignment

You are required to submit an assignment that is worth 20% (20 marks) of your final grade. This assignment will be based on research and may include an experiment. This assignment requires you to apply skills, knowledge and understanding to investigate a relevant topic in chemistry and its effect on the environment and/or society. Your school or college will provide you with a Candidate's Guide for this assignment, which has been produced by the SQA. This guide gives guidance on what is required to complete the 400–800 word report and gain as many marks as possible.

Your assignment report will be marked by the SQA.

### Component 2 - The Question Paper

The question paper will assess breadth and depth of knowledge and understanding from across all of the three Units. The question paper will require you to:

- Make statements, provide explanations, and describe information to demonstrate knowledge and understanding.
- Apply knowledge and understanding to new situations to solve problems.
- Plan and design experiments.
- Present information in various forms such as graphs, tables etc.
- Perform calculations based on information given.
- Give predictions or make generalisations based on information given.
- Draw conclusions based on information given.
- Suggest improvement to experiments to improve the accuracy of results obtained or to improve safety.

To achieve a 'C' grade in National 5 Chemistry you must achieve about 50 % of the 100 marks available when the two components, i.e. the Question Paper and the Assignment are combined. For a B you will need 60%, while for an 'A' grade you must ensure that you gain as many of the available marks as possible, and at least 70%.

This book contains model papers that cover the content of the National 5 Chemistry course and illustrate the standard, structure and requirements of the Question Paper that you will sit during your exam.

Each model paper consists of two sections. (A marking scheme for each section is provided at the end of this book.)

- Section A will contain objective questions (multiple choice) and will have 20 marks.
- Section B will contain restricted and extended response questions and will have 60 marks.

Each model paper contains a variety of questions including some that require:

- demonstration and application of knowledge, and understanding of the mandatory content of the course from across the three units
- application of scientific inquiry skills.

## How to use this book

This book can be used in two ways:

1. You can complete an entire model paper under exam conditions, without the use of books or notes, and then mark the papers using the marking scheme provided. This method gives you a clear indication of the level you are working at and should highlight the content areas that you need to work on before attempting the next model paper. This method also allows you to see your progress as you complete each model paper.

2. You can complete a model paper using your notes and books. Try the question first and then refer to your notes if you are unable to answer the question. This is a form of studying and by doing this you will cover all the areas of content that you are weakest in. You should notice that you are referring to your notes less with each model paper completed.

Try to practise as many questions as possible. This will get you used to the language used in the Question Papers and ultimately improve your chances of success.

## Some hints and tips

Below is a list of hints and tips that will help you to achieve your full potential in the National 5 exam.

- Ensure that you **read each question carefully**. Scanning the question and missing the main points results in mistakes being made. Some students highlight the main points of a question with a highlighter pen to ensure that they don't miss anything out.

- Open ended questions include the statement **'Using your knowledge of chemistry'**. These questions provide you with an opportunity to show off your chemistry knowledge. To obtain the three marks on offer for these questions, you must demonstrate a good understanding of the chemistry involved and provide a logically correct answer to the question posed.

- When doing calculations, ensure that you **show all of your working**. If you make a simple arithmetical mistake you may still be awarded some of the marks, but only if your working is laid out clearly so that the examiner can see where you went wrong and what you did correctly. Just giving the answers is very risky so you should always show your working.

- **Attempt all questions.** Giving no answer at all means that you will definitely not gain any marks.

- When you are required to read a passage to answer a question, ensure that you **read it carefully** as the information you require is contained within it. It may not be obvious at first but the answers will be contained within the passage.

- If you are asked to 'explain' in a question, then you must **explain your answer fully**. For example, if you are asked to explain how a covalent bond holds atoms together then you cannot simply say:

    'A covalent bond is a shared pair of electrons between atoms in a non-metal.

  This answer tells the examiner what a covalent bond is, but does not explain how it holds the atoms together. To gain the marks, an answer similar to this should be written:

    'A covalent bond is a shared pair of electrons between atoms in a non-metal. The shared electrons are attracted to the nuclei of both atoms, which creates a tug-of-war effect creating the covalent bond.'

- You will be required to draw one graph in each exam. To obtain all the marks, ensure that the graphs have **labels, units, points plotted correctly** and a line of 'best fit' drawn between the points.

- Use your **data booklet** when you are asked to write formulas, ionic formulas, formula mass etc. You have the data booklet in front of you so use it to double check the numbers you require.

- Work on your **timing**. The multiple-choice section (Section 1) should take approximately 30 minutes. Attempt to answer the multiple-choice questions before you look at the four possible answers, as this will improve your confidence. Use scrap paper when required to scribble down structural formulae, calculations, chemical formulae etc., as this will reduce your chance of making errors. If you are finding the question difficult, try to eliminate the obviously wrong answers to increase your chances.

- When asked to **predict or estimate** based on information from a graph or a table, then take your time to look for patterns. For example, if asked to predict a boiling point, try to establish if there is a regular change in boiling point and use that regular pattern to establish the unknown boiling point.

- When drawing a **diagram** of an experiment ask yourself the question, 'Would this work if I set it up exactly like this in the lab?' Ensure that the method you have drawn would produce the desired results safely. If, for example, you are heating a flammable reactant such as alcohol then you will not gain the marks if you heat it with a Bunsen burner in your diagram; a water bath would be much safer! Make sure your diagram is labeled clearly.

## Good luck!

Remember that the rewards for passing National 5 Chemistry are well worth it! Your pass will help you get the future you want for yourself. In the exam, be confident in your own ability. If you're not sure how to answer a question, trust your instincts and just give it a go anyway. Keep calm and don't panic! GOOD LUCK!

# 2013 Specimen Question Paper

# National Qualifications
## SPECIMEN ONLY

**SQO6/N5/01**

# Chemistry
## Section 1—Questions

Date — Not applicable

Duration — 2 hours

Instructions for completion of Section 1 are given on Page two of the question paper SQO6/N5/02.

Record your answers on the grid on Page three of your answer booklet.

Do NOT write in this booklet.

Before leaving the examination room you must give your answer booklet to the Invigilator. If you do not, you may lose all the marks for this paper.

## SECTION 1

1. Which of the following elements exists as a covalent network?

   A  Helium

   B  Nitrogen

   C  Silicon

   D  Sulfur

2. Which line in the table correctly describes an electron?

   |   | Mass | Charge |
   |---|------|--------|
   | A | negligible | +1 |
   | B | negligible | −1 |
   | C | 1 | +1 |
   | D | 1 | 0 |

3. Solid ionic compounds do **not** conduct electricity because

   A  the ions are not free to move

   B  the electrons are not free to move

   C  solid substances never conduct electricity

   D  there are no charged particles in ionic compounds.

4. The shapes of some molecules are shown below.

   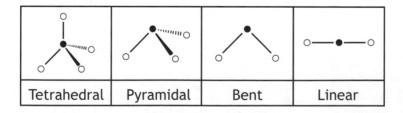

   | Tetrahedral | Pyramidal | Bent | Linear |
   |---|---|---|---|

   Phosphine is a compound of phosphorus and hydrogen. The shape of a molecule of phosphine is likely to be

   A  tetrahedral

   B  pyramidal

   C  bent

   D  linear.

Questions **5** and **6** refer to the table below.

The table shows information about some particles.

| | Number of | | |
| --- | --- | --- | --- |
| Particle | protons | neutrons | electrons |
| A | 9 | 10 | 10 |
| B | 11 | 12 | 11 |
| C | 15 | 16 | 15 |
| D | 19 | 20 | 18 |

5. Identify the particle which is a negative ion.

6. Identify the particle which would give a lilac flame colour.

   *You may wish to use the data booklet to help you.*

7. Which of the following statements correctly describes the concentrations of $H^+(aq)$ and $OH^-(aq)$ ions in pure water?

   A   The concentrations of $H^+(aq)$ and $OH^-(aq)$ ions are equal.

   B   The concentrations of $H^+(aq)$ and $OH^-(aq)$ ions are zero.

   C   The concentration of $H^+(aq)$ ions is greater than the concentration of $OH^-(aq)$ ions.

   D   The concentration of $H^+(aq)$ ions is less than the concentration of $OH^-(aq)$ ions.

8.

   The name of the above compound is

   A   2-ethylpropane

   B   1,1-dimethylpropane

   C   2-methylbutane

   D   3-methylbutane.

9. Which of the following could be the molecular formula of a cycloalkane?

   A　$C_6H_8$

   B　$C_6H_{10}$

   C　$C_6H_{12}$

   D　$C_6H_{14}$

10. In which of the following reactions is oxygen used up?

    A　Combustion

    B　Neutralisation

    C　Addition

    D　Polymerisation

11. Which line in the table correctly shows the two families of compounds which react together to produce esters?

| A | carboxylic acid | cycloalkane |
| B | alcohol | alkene |
| C | cycloalkane | alkene |
| D | carboxylic acid | alcohol |

**12.** Which of the following molecules is an isomer of hept-2-ene?

A

```
     H    H    H    H    H
     |    |    |    |    |
H —  C — C — C — C — C — H
     |    |    |    |    |
     H    H    H    |    H
                    H — C — H
                        |
                        H
```

B

```
     H    H    H    H    H    H
     |    |    |    |    |    |
H —  C — C — C — C — C — C — H
     |    |    |    |    |    |
     H    H    H    |    H    H
                    H — C — H
                        |
                        H
```

C

```
                         H
                         |
                    H — C — H
     H    H    H    H    |    H
     |    |    |    |         |
H —  C — C — C — C — C = C — H
     |    |    |    |
     H    H    H    H
```

D

```
     H    H    H    H    H    H    H
     |    |    |    |    |    |    |
H —  C — C — C — C — C = C — C — H
     |    |    |    |              |
     H    H    H    H              H
```

**13.** A student tested some compounds.  The results are given in the table.

| Compound | pH of aqueous solution | Effect on bromine solution |
|---|---|---|
| (structure: propanoic acid) | 4 | no effect |
| (structure: propenoic acid) | 4 | decolourised |
| (structure: propan-1-ol) | 7 | no effect |
| (structure: propen-1-ol) | 7 | decolourised |

Which line in the table below shows the correct results for the following compound?

| | pH of aqueous solution | Effect on bromine solution |
|---|---|---|
| A | 4 | decolourised |
| B | 7 | decolourised |
| C | 4 | no effect |
| D | 7 | no effect |

14. Which of the following diagrams could be used to represent the structure of a metal?

A

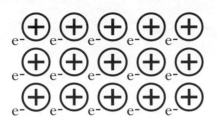

B

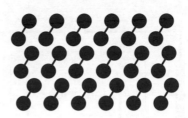

C

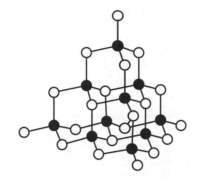

D

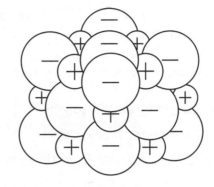

15. Which of the following metals does **not** react with dilute acid?

    A    Magnesium

    B    Calcium

    C    Copper

    D    Zinc

16. Which of the following metals can be extracted from its oxide by heat alone?

    A    Aluminium

    B    Iron

    C    Silver

    D    Zinc

**17.**

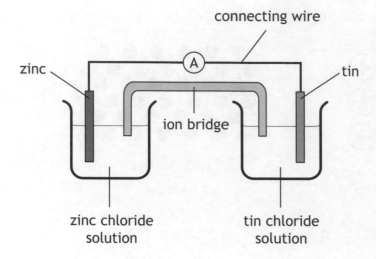

In the cell shown, electrons flow through

A   the solution from tin to zinc

B   the solution from zinc to tin

C   the connecting wire from tin to zinc

D   the connecting wire from zinc to tin.

**18.**   Four cells were made by joining copper, iron, tin and zinc to silver.

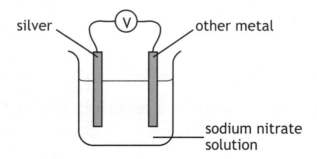

The voltages are shown in the table.

Which line in the table below shows the voltage of the cell containing copper joined to silver?

*You may wish to use the data booklet to help you.*

| Cell | Voltage (V) |
|------|-------------|
| A    | 1·6         |
| B    | 1·2         |
| C    | 0·9         |
| D    | 0·5         |

19. The ion-electron equation for the oxidation and reduction steps in the reaction between magnesium and silver(I) ions are:

$Mg \rightarrow Mg^{2+} + 2e^-$

$Ag^+ + e^- \rightarrow Ag$

The overall redox equation is

A $\quad Mg + 2Ag^+ \rightarrow Mg^{2+} + 2Ag$

B $\quad Mg + Ag^+ \rightarrow Mg^{2+} + Ag$

C $\quad Mg + Ag^+ + e^- \rightarrow Mg^{2+} + Ag + 2e^-$

D $\quad Mg + 2Ag \rightarrow Mg^{2+} + 2Ag^+$.

20. The structure below shows a section of an addition polymer.

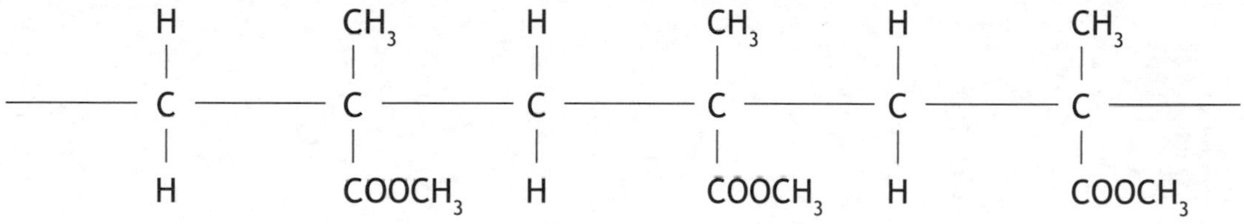

Which of the following molecules is used to make this polymer?

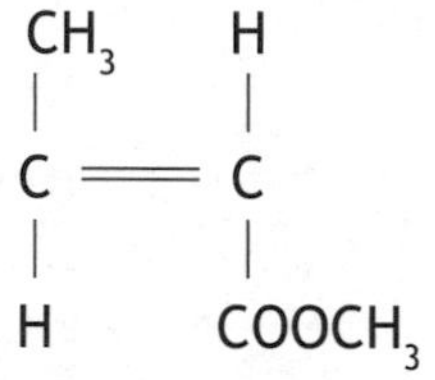

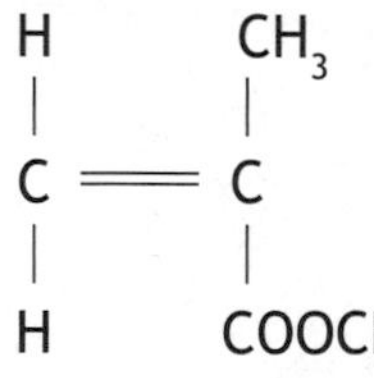

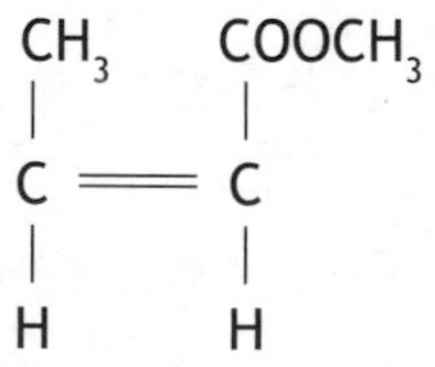

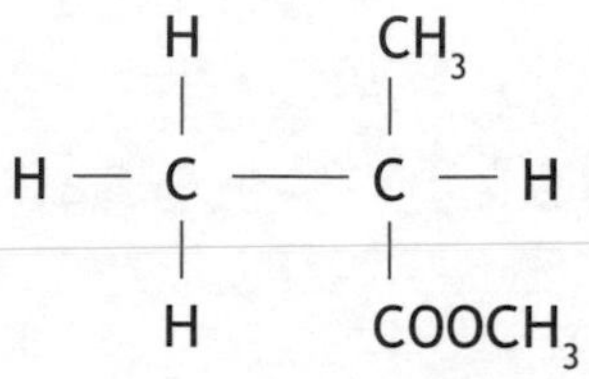

[END OF SECTION 1. NOW ATTEMPT THE QUESTIONS IN SECTION 2 OF YOUR QUESTION AND ANSWER BOOKLET.]

# N5

Mark

**National Qualifications**
**SPECIMEN ONLY**

**SQ06/N5/02**

**Chemistry**
**Section 1—Answer**
**Grid and Section 2**

Date — Not applicable

Duration — 2 hours

**Fill in these boxes and read what is printed below.**

Full name of centre

Town

Forename(s)

Surname

Number of seat

Date of birth

Day    Month    Year

D D    M M    Y Y

Scottish candidate number

**Total marks — 80**

**SECTION 1 — 20 marks**

Attempt ALL questions in this section.

Instructions for completion of Section 1 are given on Page two.

**SECTION 2 — 60 marks**

Attempt ALL questions in this section.

Read all questions carefully before attempting.

Use **blue** or **black** ink. Do NOT use gel pens.

Write your answers in the spaces provided. Additional space for answers and rough work is provided at the end of this booklet. If you use this space, write clearly the number of the question you are attempting. Any rough work must be written in this booklet. You should score through your rough work when you have written your fair copy.

Before leaving the examination room you must give this booklet to the Invigilator. If you do not, you may lose all the marks for this paper.

✕ SQA

**SECTION 1— 20 marks**

The questions for Section 1 are contained in the booklet Chemistry Section 1—Questions. Read these and record your answers on the grid on Page three opposite.

1. The answer to each question is **either** A, B, C or D. Decide what your answer is, then fill in the appropriate bubble (see sample question below).

2. There is **only one correct** answer to each question.

3. Any rough working should be done on the additional space for rough working and answers sheet.

**Sample Question**

To show that the ink in a ball-pen consists of a mixture of dyes, the method of separation would be:

    A    fractional distillation

    B    chromatography

    C    fractional crystallisation

    D    filtration.

The correct answer is **B**—chromatography. The answer **B** bubble has been clearly filled in (see below).

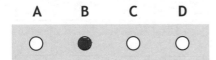

**Changing an answer**

If you decide to change your answer, cancel your first answer by putting a cross through it (see below) and fill in the answer you want. The answer below has been changed to **D**.

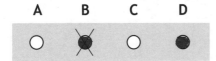

If you then decide to change back to an answer you have already scored out, put a tick (✓) to the **right** of the answer you want, as shown below:

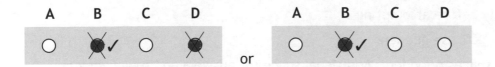

*Page two*

## SECTION 1—Answer Grid

|    | A | B | C | D |
|----|---|---|---|---|
| 1  | ○ | ○ | ○ | ○ |
| 2  | ○ | ○ | ○ | ○ |
| 3  | ○ | ○ | ○ | ○ |
| 4  | ○ | ○ | ○ | ○ |
| 5  | ○ | ○ | ○ | ○ |
| 6  | ○ | ○ | ○ | ○ |
| 7  | ○ | ○ | ○ | ○ |
| 8  | ○ | ○ | ○ | ○ |
| 9  | ○ | ○ | ○ | ○ |
| 10 | ○ | ○ | ○ | ○ |
| 11 | ○ | ○ | ○ | ○ |
| 12 | ○ | ○ | ○ | ○ |
| 13 | ○ | ○ | ○ | ○ |
| 14 | ○ | ○ | ○ | ○ |
| 15 | ○ | ○ | ○ | ○ |
| 16 | ○ | ○ | ○ | ○ |
| 17 | ○ | ○ | ○ | ○ |
| 18 | ○ | ○ | ○ | ○ |
| 19 | ○ | ○ | ○ | ○ |
| 20 | ○ | ○ | ○ | ○ |

[BLANK PAGE]

**MARKS**

## SECTION 2— 60 marks

## Attempt ALL questions.

1.  Graphs can be used to show the change in the rate of a reaction as the reaction proceeds.

    The graph shows the volume of gas produced in an experiment over a period of time.

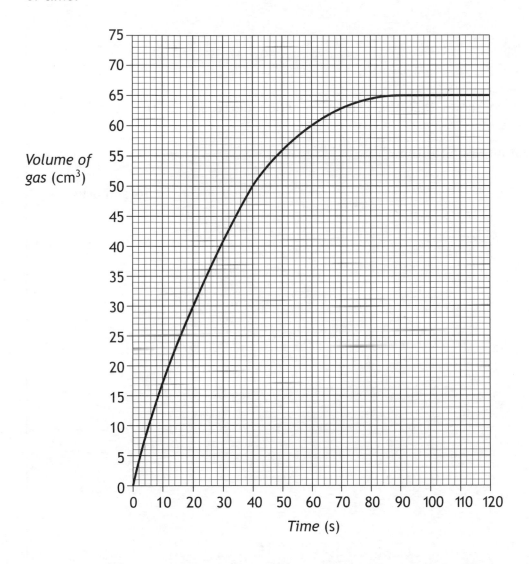

    (a)  State the time, in seconds, at which the reaction stopped.     1

MARKS | DO NOT WRITE IN THIS MARGIN

**1. (continued)**

(b) Calculate the average rate of reaction, in $cm^3 s^{-1}$, for the first 20 seconds.

**Show your working clearly.**

2

(c) The graph shows that the rate of reaction decreases as the reaction proceeds.

Suggest a reason for this.

1

**Total marks    4**

MARKS | DO NOT WRITE IN THIS MARGIN

**2.** The group 7 element bromine was discovered by Balard in 1826.

Bromine gets its name from the Greek 'bromos' meaning stench.

Bromine consists of a mixture of two isotopes, $^{79}_{35}$Br and $^{81}_{35}$Br.

(a) What is meant by the term isotope? 1

(b) Complete the table for $^{79}_{35}$Br. 1

| Isotope | Number of protons | Number of neutrons |
|---------|-------------------|--------------------|
| $^{79}_{35}$Br | | |

(c) The relative atomic mass of an element can be calculated using the formula:

$$\frac{(\text{mass of isotope A} \times \% \text{ of isotope A}) + (\text{mass of isotope B} \times \% \text{ of isotope B})}{100}$$

A sample of bromine contains 55% of the isotope with mass 79 and 45% of the isotope with mass 81.

Calculate the relative atomic mass of bromine in this sample. 2

**Show your working clearly.**

MARKS | DO NOT WRITE IN THIS MARGIN

**2. (continued)**

(d) In 1825 bromine had been isolated from sea water by Liebig who mistakenly thought it was a compound of iodine and chlorine.

Using your knowledge of chemistry, comment on why Liebig might have made this mistake. 3

**Total marks 7**

MARKS | DO NOT WRITE IN THIS MARGIN

3. (a) Sulfur dioxide gas is produced when fossil fuels containing sulfur are burned.

When sulfur dioxide dissolves in water in the atmosphere "acid rain" is produced.

Circle the correct phrase to complete the sentence.

1

Compared with pure water, acid rain contains $\begin{cases} \text{a higher} \\ \text{a lower} \\ \text{the same} \end{cases}$ concentration of hydrogen ions.

(b) The table shows information about the solubility of sulfur dioxide.

| Temperature (°C) | 0 | 20 | 30 | 40 | 50 | 60 |
|---|---|---|---|---|---|---|
| Solubility (g/100cm³) | 22·0 | 10·0 | 6·0 | 3·0 | 2·0 | 1·5 |

(i) Draw a line graph of solubility against temperature.

Use appropriate scales to fill most of the graph paper.

3

(Additional graph paper, if required, will be found on *Page twenty-seven*.)

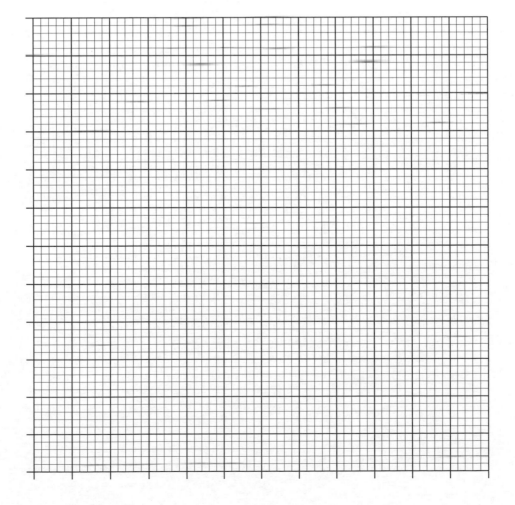

MARKS DO NOT WRITE IN THIS MARGIN

**3. (b) (continued)**

(ii) Using your graph, estimate the solubility of sulfur dioxide, in $g/100\ cm^3$, at 10 °C. 1

**Total marks 5**

MARKS | DO NOT WRITE IN THIS MARGIN

4.  A student investigated the reaction of carbonates with dilute hydrochloric acid.

(a)  In one reaction lithium carbonate reacted with dilute hydrochloric acid.

The equation for the reaction is:

$$Li_2CO_3(s) \ + \ HCl(aq) \ \rightarrow \ LiCl(aq) \ + \ CO_2(g) \ + \ H_2O(\ell)$$

(i)  Balance this equation.    **1**

(ii)  Identify the salt produced in this reaction.    **1**

(b)  In another reaction 1 g of calcium carbonate reacted with excess dilute hydrochloric acid.

$$CaCO_3(s) \ + \ 2HCl(aq) \ \rightarrow \ CaCl_2(aq) \ + \ CO_2(g) \ + \ H_2O(\ell)$$

(i)  Calculate the mass, in grams, of carbon dioxide produced.    **3**

MARKS | DO NOT WRITE IN THIS MARGIN

**4. (b) (continued)**

(ii) The student considered two methods to confirm the mass of carbon dioxide gas produced in this reaction.

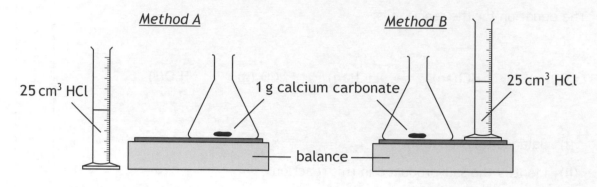

_Method A_                    _Method B_

25 cm³ HCl    1 g calcium carbonate    25 cm³ HCl

balance

| Method A | Method B |
|---|---|
| 1. Add the acid from the measuring cylinder to the calcium carbonate in the flask. | 1. Weigh the flask with the calcium carbonate and the acid in the measuring cylinder together. |
| 2. Weigh the flask and contents. | 2. Add the acid from the measuring cylinder to the calcium carbonate in the flask and replace the empty measuring cylinder on the balance. |
| 3. Leave until no more bubbles are produced. | 3. Leave until no more bubbles are produced. |
| 4. Reweigh the flask and contents. | 4. Reweigh the flask, contents and the empty measuring cylinder together. |

Explain which method would give a more reliable estimate of the mass of carbon dioxide produced during the reaction.    2

**Total marks    7**

MARKS | DO NOT WRITE IN THIS MARGIN

5. Antacid tablets are used to treat indigestion which is caused by excess acid in the stomach.

Different brands of tablets contain different active ingredients.

| Name of active ingredient | magnesium carbonate | calcium carbonate | magnesium hydroxide | aluminium hydroxide |
|---|---|---|---|---|
| Reaction with acid | fizzes | fizzes | does not fizz | does not fizz |
| Cost per gram (pence) | 16·0 | 11·0 | 7·5 | 22·0 |
| Mass of solid needed to neutralise 20 cm$^3$ of acid (g) | 0·7 | 1·2 | 0·6 | 0·4 |
| Cost of neutralising 20 cm$^3$ of acid (pence) | | 13·2 | 4·5 | 8·8 |

(a) Write the **ionic** formula for aluminium hydroxide.

1

(b) (i) Complete the table to show the cost of using magnesium carbonate to neutralise 20 cm$^3$ of acid.

1

(ii) Which **one** of the four active ingredients would **you** use to neutralise excess stomach acid?

Explain your choice.

1

**Total marks    3**

MARKS | DO NOT WRITE IN THIS MARGIN

6.  Read the passage below and answer the questions that follow.

---

**Potassium Permanganate (KMnO$_4$)—The Purple Solution**

Potassium permanganate's strong oxidising properties make it an effective disinfectant. Complaints such as athlete's foot and some fungal infections are treated by bathing the affected area in KMnO$_4$ solution.

In warm climates vegetables are washed in KMnO$_4$ to kill bacteria such as E. coli and S. aureus. Chemists use KMnO$_4$ in the manufacture of saccharin, ascorbic acid (vitamin C) and benzoic acid.

Baeyer's reagent is an alkaline solution of KMnO$_4$ and is used to detect unsaturated organic compounds. The reaction of KMnO$_4$ with alkenes is also used to extend the shelf life of fruit. Ripening fruit releases ethene gas which causes other fruit to ripen. Shipping containers are fitted with gas scrubbers that use alumina or zeolite impregnated with KMnO$_4$ to stop the fruit ripening too quickly.

$$C_2H_4 + 4KMnO_4 \rightarrow 4MnO_2 + 4KOH + 2CO_2$$

The scrubbers indicate when they need to be replaced because the purple colour changes to brown as the KMnO$_4$ is used up.

---

The passage on potassium permanganate was taken from an article by Simon Cotton on "Soundbite molecules" in "Education in Chemistry" November 2009.

(a)  Suggest a pH for Baeyer's reagent.          1

(b)  Name the gas removed by the scrubbers.          1

(c)  Name a chemical mentioned in the passage which contains the following functional group.          1

                                        **Total marks  3**

**MARKS**
DO NOT WRITE IN THIS MARGIN

7. In the 2012 London Olympics, alkanes were used as fuels for the Olympic flame.

   (a) The torches that carried the Olympic flame across Britain burned a mixture of propane and butane.

   Propane and butane are members of the same homologous series.

   What is meant by the term homologous series?    1

   (b) Natural gas, which is mainly methane, was used to fuel the flame in the Olympic cauldron.

   (i) Draw a diagram to show how **all** the outer electrons are arranged in a molecule of methane, $CH_4$.    1

MARKS | DO NOT WRITE IN THIS MARGIN

**7. (b) (continued)**

    **(ii)** Methane is a covalent molecular substance.

        It has a low boiling point and is a gas at room temperature.

        Explain why methane is a gas at room temperature.        **1**

                                        **Total marks    3**

MARKS | DO NOT WRITE IN THIS MARGIN

8. Car manufacturers have developed flexible fuel engines for vehicles. These vehicles can run on ethanol or petrol or a mixture of both.

   Ethanol can be produced from ethene which comes from cracking crude oil. It can also be made by fermenting glucose which is obtained from crops such as sugar cane and maize.

   (a) The structure of ethanol is shown below.

$$
\begin{array}{c}
\quad\ \ \ H\ \ \ \ H \\
\quad\ \ \ | \quad\ \ | \\
H - C - C - O - H \\
\quad\ \ \ | \quad\ \ | \\
\quad\ \ \ H\ \ \ \ H
\end{array}
$$

   Circle the functional group in this molecule.          1

   (b) Ethanol is produced from ethene as shown.

$$
\begin{array}{c}
H \qquad\qquad H \\
\ \backslash \qquad\quad\ / \\
\ \ C = C \qquad + \quad H_2O \quad \longrightarrow \\
\ / \qquad\quad\ \backslash \\
H \qquad\qquad H
\end{array}
\qquad
\begin{array}{c}
H\ \ \ \ H \\
| \quad\ \ | \\
H - C - C - O - H \\
| \quad\ \ | \\
H\ \ \ \ H
\end{array}
$$

          ethene                              ethanol

   (i) Name the **type** of chemical reaction taking place.          1

   (ii) Draw a structural formula for a product of the following reaction.          1

$$
\begin{array}{c}
\ \ \ \ \ H\ \ \ \ H\ \ \ \ H\ \ \ \ H \\
\ \ \ \ \ | \quad\ | \quad\ | \quad\ | \\
H - C - C - C = C \quad + \quad H_2O \\
\ \ \ \ \ | \quad\ | \qquad\quad | \\
\ \ \ \ \ H \quad\ | \qquad\quad H \\
\ \ \ \ \ \ \ \ \ \ H - C - H \\
\ \ \ \ \ \ \ \ \ \ \ \ \ \ \ | \\
\ \ \ \ \ \ \ \ \ \ \ \ \ \ \ H
\end{array}
\qquad \downarrow
$$

MARKS | DO NOT WRITE IN THIS MARGIN

**8. (continued)**

(c) Suggest **one** disadvantage of producing ethanol from crops. 1

(d) Ethanol can be used to produce ethanoic acid.

(i) Draw a structural formula for ethanoic acid. 1

(ii) To which family of compounds does ethanoic acid belong? 1

**Total marks 6**

MARKS | DO NOT WRITE IN THIS MARGIN

**9.** Alkanes burn, releasing energy.

(a) What name is given to any chemical reaction which releases energy?

1

(b) A student investigated the amount of energy released when an alkane burns using the apparatus shown.

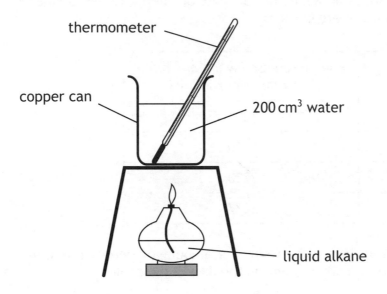

thermometer

copper can

200 cm³ water

liquid alkane

The student recorded the following data.

| Mass of alkane burned | 1 g |
|---|---|
| Volume of water | 200 cm³ |
| Initial temperature of water | 15 °C |
| Final temperature of water | 55 °C |
| Specific heat capacity of water | 4·18 kJ kg⁻¹ °C⁻¹ |

(i) Calculate the energy released, in kJ.

*You may wish to use the data booklet to help you.*
**Show your working clearly.**

3

MARKS DO NOT WRITE IN THIS MARGIN

**9. (b) (continued)**

(ii) Suggest **one** improvement to the student's investigation. 1

(c) The table gives information about the amount of energy released when one mole of some alkanes are burned.

| *Name of alkane* | *Energy released when one mole of alkane is burned* (kJ) |
|---|---|
| methane | 891 |
| ethane | 1560 |
| propane | 2220 |
| butane | 2877 |

(i) Describe the relationship between the amount of energy released and the number of carbon atoms in the alkane molecule. 1

(ii) Predict the amount of heat released, in kJ, when one mole of pentane is burned. 1

**Total marks 7**

MARKS DO NOT WRITE IN THIS MARGIN

**10.** The essential elements for plant growth are nitrogen, phosphorus and potassium.

A student was asked to prepare a dry sample of a compound which contained **two** of these elements.

The student was given access to laboratory equipment and the following chemicals.

| *Chemical* | *Formula* |
|---|---|
| ammonium hydroxide | $NH_4OH$ |
| magnesium nitrate | $Mg(NO_3)_2$ |
| nitric acid | $HNO_3$ |
| phosphoric acid | $H_3PO_4$ |
| potassium carbonate | $K_2CO_3$ |
| potassium hydroxide | $KOH$ |
| sodium hydroxide | $NaOH$ |
| sulfuric acid | $H_2SO_4$ |
| water | $H_2O$ |

Using your knowledge of chemistry, comment on how the student could prepare their dry sample. **3**

*You may wish to use the data booklet to help you.*

11.   Urea, $H_2NCONH_2$, can be used as a fertiliser.

(a)  Calculate the percentage of nitrogen in urea.      3
     **Show your working clearly.**

(b)  Other nitrogen based fertilisers can be produced from ammonia.

     In industry, ammonia is produced in the Haber process using a catalyst.

$$N_2(g) + 3H_2(g) \rightleftharpoons 2NH_3(g)$$

Suggest why a catalyst may be used in an industrial process.      1

**Total marks**    4

**MARKS** | DO NOT WRITE IN THIS MARGIN

**12.** Technetium-99m is used in medicine to detect damage to heart tissue.

It is a gamma-emitting radioisotope and is injected into the body.

(a) The half-life of technetium-99m is 6 hours.

How much of a 2 g sample of technetium-99m would be left after 12 hours?    **2**

(b) Suggest one reason why technetium-99m can be used safely in this way.    **1**

(c) Technetium-99m is formed when molybdenum-99 decays.

The decay equation is:

$$^{99}_{42}\text{Mo} \rightarrow {}^{99}_{43}\text{Tc} + \text{X}$$

Identify **X**.    **1**

**Total Marks   4**

MARKS | DO NOT WRITE IN THIS MARGIN

13. The concentration of chloride ions in water affects the ability of some plants to grow.

A student investigated the concentration of chloride ions in the water at various points along the river Tay.

The concentration of chloride ions in water can be determined by reacting the chloride ions with silver ions.

$$Ag^+(aq) \; + \; Cl^-(aq) \; \rightarrow \; AgCl(s)$$

A $20 \, cm^3$ water sample gave a precipitate of silver chloride with a mass of $1 \cdot 435 \, g$.

(a) Calculate the number of moles of silver chloride, AgCl, present in this sample.

**Show your working clearly.**

2

(b) Using your answer to part (a), calculate the concentration, in $mol \, l^{-1}$, of chloride ions in this sample.

**Show your working clearly.**

2

Total marks    4

[END OF SPECIMEN QUESTION PAPER]

**MARKS** | DO NOT WRITE IN THIS MARGIN

**ADDITIONAL SPACE FOR ROUGH WORKING AND ANSWERS**

**ADDITIONAL SPACE FOR ROUGH WORKING AND ANSWERS**

**ADDITIONAL SPACE FOR ANSWERS**

Additional graph paper for Question 3 (b) (i)

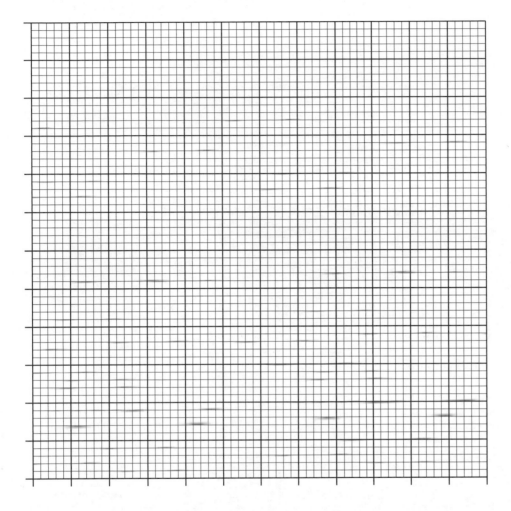

NATIONAL 5

# 2013 Model Paper 1

**N5** National
Qualifications
MODEL PAPER 1

# Chemistry
# Section 1—Questions

Duration — 2 hours

Instructions for completion of Section 1 are given on Page two of the question paper.

Record your answers on the grid on Page three of your answer booklet.

Do not write in this booklet.

Before leaving the examination room you must give your answer booklet to the Invigilator. If you do not, you may lose all the marks for this paper.

HODDER
GIBSON
LEARN MORE

## SECTION 1

1.  Which of the following elements has a covalent molecular structure?

    A    Sodium

    B    Helium

    C    Silicon

    D    Hydrogen.

2.  An element has an atomic number of 11 and a mass number of 23. The number of electrons present in an atom of this element is?

    A    11

    B    12

    C    23

    D    34

3.  Ionic compounds conduct in solution because

    A    the ions are free to move

    B    the ions are not free to move

    C    the electrons are free to move

    D    the electrons are not free to move.

4.  Which of the following compounds has molecules with the same shape as ammonia?

    A    Carbon dioxide

    B    Hydrogen oxide

    C    Sulfur dioxide

    D    Phosphorus hydride.

5.  Which line in the table correctly describes a neutron?

    |   | Mass | Charge |
    |---|------|--------|
    | A | 1 | −1 |
    | B | negligible | 0 |
    | C | 1 | +1 |
    | D | 1 | 0 |

Questions 6 and 7 refer to the table below.

The table contains information about some substances.

| Substance | Melting point ( ) $^0C$ | Boiling point ( ) $^0C$ | Conducts as a solid | Conducts as a liquid |
|-----------|-----------|-----------|-----------|-----------|
| A | -7 | 59 | No | No |
| B | 1492 | 2897 | Yes | Yes |
| C | 1407 | 2357 | No | No |
| D | 606 | 1305 | No | Yes |

6. Identify the substance that is a liquid at room temperature (21 °C).

7. Identify the substance that exists as a covalent network.

8. Which of the following statements correctly describes the concentration of $H^+(aq)$ and OH (aq) ions in an acidic substance when compared to pure water?

   A   The concentration of $H^+(aq)$ and $OH^-(aq)$ ions are zero.

   B   The concentration of $H^+(aq)$ and $OH^-(aq)$ ions are equal.

   C   The concentration of $H^+(aq)$ is higher than $OH^-(aq)$ ions.

   D   The concentration of $H^+(aq)$ is lower than $OH^-(aq)$ ions.

9.

```
    H   H           H
    |   |           |
H — C — C — C = C — C — H
    |   |   |   |   |
    H   H   H   H   H
```

   The name of the above compound is

   A   but-2-ene

   B   pent-2-ene

   C   but-3-ene

   D   pent-3-ene.

10. Shown below is the structure of a compound known as neopentane.

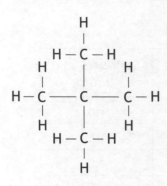

The systematic name of the above compound is

A   2,3-dimethylbutane

B   3,2-dimethylbutane

C   2,2-dimethylpropane

D   3,2-dimethylpropane.

11. Which of the following structures belongs to the same homologous series as the compound with the formula $C_3H_8$?

12. Which of the following compounds is **not** an isomer of pent-1-ene?

A   but-1-ene

B   pent-2-ene

C   cyclopentane

D   2-methylbut-1-ene

**13.** Identify the hydrocarbon that reacts quickly with bromine solution.

A

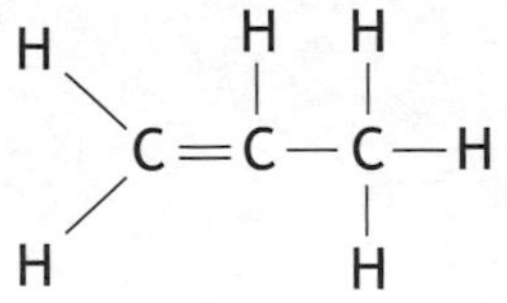

B

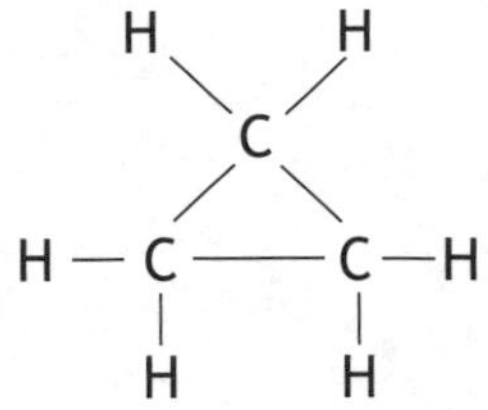

C

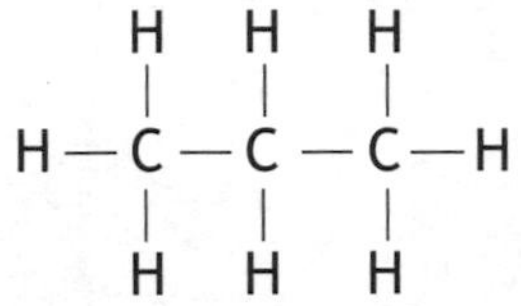

D

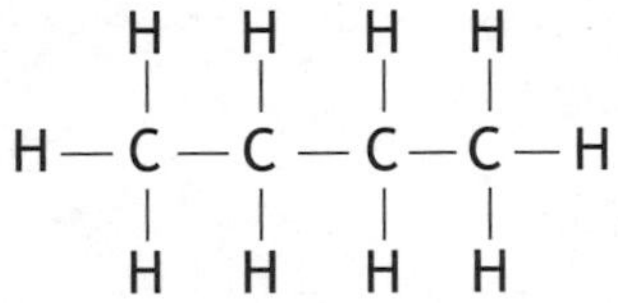

**14.** Metallic bonds are due to

A   a shared pair of electrons

B   an attraction between positive ions and negative ions

C   an attraction between positive ions and delocalised electrons

D   an attraction between negative ions and delocalised electrons.

**15.** $Cu^{2+} + 2e^- \rightarrow Cu$

This ion electron equation represents the

A   reduction of copper(II) ions

B   reduction of copper(I) ions

C   oxidation of copper(II) ions

D   oxidation of copper(I) ions.

**16.** Which of the following metals can be obtained from its ore by heat alone?

A   Iron

B   Potassium

C   Mercury

D   Aluminium

**17.**

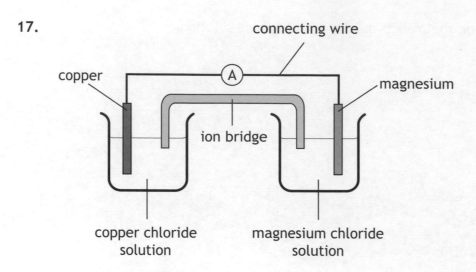

In the cell shown, electrons flow through

A    the solution from copper to magnesium

B    the solution from magnesium to copper

C    the connecting wire from copper to magnesium

D    the connecting wire from magnesium to copper.

**18.** Four cells were made by joining magnesium, aluminium, nickel and zinc to lead. The voltages are shown in the table.

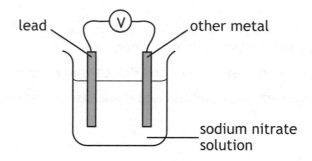

Which line in the table below shows the voltage of the cell containing magnesium joined to lead?

*You may wish to use the data booklet to help you.*

| Cell | Voltage (V) |
|------|-------------|
| A    | 0·3         |
| B    | 1·0         |
| C    | 1·1         |
| D    | 1·2         |

19. The half-life of the isotope $^{210}Pb$ is 22 years.

What fraction of the original sample will remain after 44 years?

A    $\frac{1}{2}$

B    $\frac{1}{4}$

C    $\frac{1}{8}$

D    $\frac{1}{16}$

20. The structure shows a section of the addition polymer Teflon.

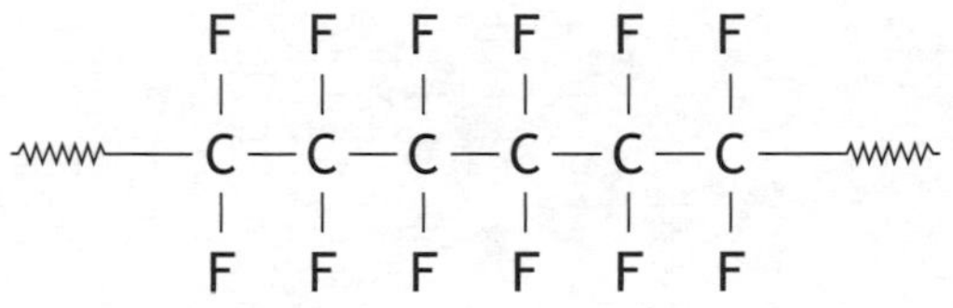

Which molecule is used to make this polymer?

A

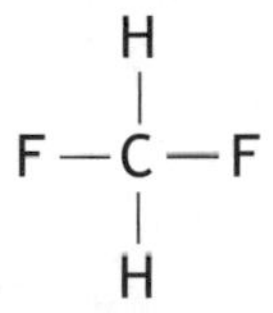

B

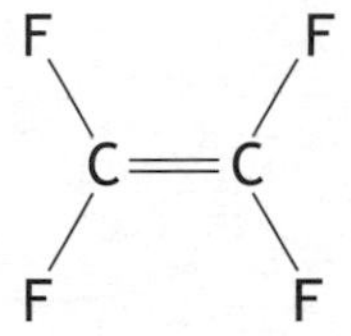

C

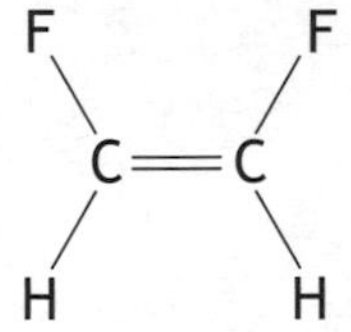

D

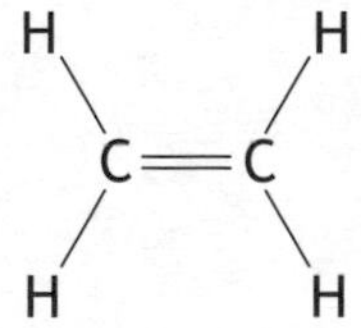

**[END OF SECTION 1. NOW ATTEMPT THE QUESTIONS IN SECTION 2 OF YOUR QUESTION AND ANSWER BOOKLET.]**

**N5** National Qualifications
MODEL PAPER 1

# Chemistry
## Section 1—Answer Grid and Section 2

Duration — 2 hours

**Total marks — 80**

**SECTION 1 — 20 marks**

Attempt ALL questions in this section.

Instructions for completion of Section 1 are given on Page two.

**SECTION 2 — 60 marks**

Attempt ALL questions in this section.

Read all questions carefully before attempting.

Use **blue** or **black** ink. Do NOT use gel pens.

Write your answers in the spaces provided. Additional space for answers and rough work is provided at the end of this booklet. If you use this space, write clearly the number of the question you are attempting. Any rough work must be written in this booklet. You should score through your rough work when you have written your fair copy.

HODDER GIBSON
LEARN MORE

## SECTION 1— 20 marks

The questions for Section 1 are contained in the booklet Chemistry Section 1—Questions.
Read these and record your answers on the grid on Page three opposite.

1.   The answer to each question is **either** A, B, C or D.  Decide what your answer is, then fill in the appropriate bubble (see sample question below).

2.   There is **only one correct** answer to each question.

3.   Any rough working should be done on the additional space for rough working and answers sheet.

### Sample Question

To show that the ink in a ball-pen consists of a mixture of dyes, the method of separation would be:

    A    fractional distillation

    B    chromatography

    C    fractional crystallisation

    D    filtration.

The correct answer is **B**—chromatography.   The answer **B** bubble has been clearly filled in (see below).

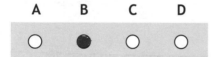

### Changing an answer

If you decide to change your answer, cancel your first answer by putting a cross through it (see below) and fill in the answer you want. The answer below has been changed to **D**.

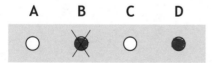

If you then decide to change back to an answer you have already scored out, put a tick (✓) to the **right** of the answer you want, as shown below:

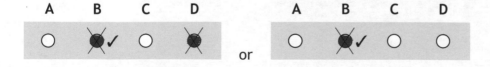

                             or

## SECTION 1—Answer Grid

| | A | B | C | D |
|---|---|---|---|---|
| 1 | ○ | ○ | ○ | ○ |
| 2 | ○ | ○ | ○ | ○ |
| 3 | ○ | ○ | ○ | ○ |
| 4 | ○ | ○ | ○ | ○ |
| 5 | ○ | ○ | ○ | ○ |
| 6 | ○ | ○ | ○ | ○ |
| 7 | ○ | ○ | ○ | ○ |
| 8 | ○ | ○ | ○ | ○ |
| 9 | ○ | ○ | ○ | ○ |
| 10 | ○ | ○ | ○ | ○ |
| 11 | ○ | ○ | ○ | ○ |
| 12 | ○ | ○ | ○ | ○ |
| 13 | ○ | ○ | ○ | ○ |
| 14 | ○ | ○ | ○ | ○ |
| 15 | ○ | ○ | ○ | ○ |
| 16 | ○ | ○ | ○ | ○ |
| 17 | ○ | ○ | ○ | ○ |
| 18 | ○ | ○ | ○ | ○ |
| 19 | ○ | ○ | ○ | ○ |
| 20 | ○ | ○ | ○ | ○ |

[BLANK PAGE]

**MARKS** | DO NOT WRITE IN THIS MARGIN

## SECTION 2— 60 marks

## Attempt ALL questions.

1. Rapid inflation of airbags in cars is caused by the production of nitrogen gas.

   The graph gives information on the volume of gas produced over 30 microseconds.

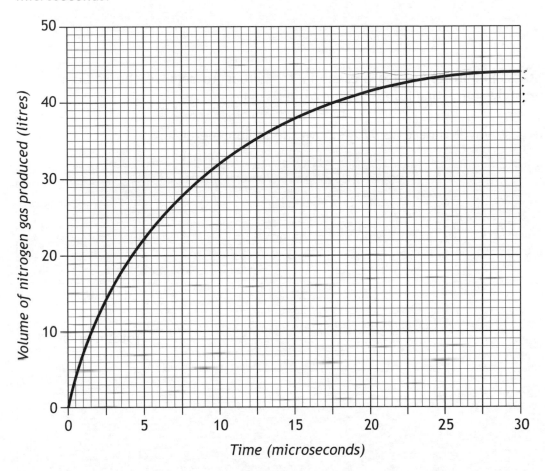

   (a)  (i)  Calculate the average rate of reaction, in litres per microsecond, between 2 and 10 microseconds.    **2**

        (ii)  At what time, in microseconds, has half of the final volume of nitrogen gas been produced?    **1**

MARKS | DO NOT WRITE IN THIS MARGIN

**1. (continued)**

(b) In some types of airbag, electrical energy causes sodium azide, $NaN_3$, to decompose producing sodium metal and nitrogen gas.

Using symbols and formulae write an equation for this reaction.    **1**
*There is no need to balance this equation.*

(c) Potassium nitrate is also present in the airbag to remove the sodium metal by converting it into sodium oxide.

Suggest why it is necessary to remove the sodium metal.    **1**

**Total marks    5**

MARKS DO NOT WRITE IN THIS MARGIN

2. Fuels have developed greatly in the past 200 years. More traditional fuels such as candle wax, peat and coal were the most commonly used fuels in the 19th century.

Using your knowledge of chemistry, describe how you could establish which of these fuels was the most efficient and which produced the least pollution. 3

**Total marks 3**

MARKS DO NOT WRITE IN THIS MARGIN

3.  Egg shells are made up mainly of calcium carbonate. A pupil carried out an experiment to react egg shell with dilute hydrochloric acid. A gas was produced.

(a)  Complete the diagram to show the apparatus which could have been used to collect and measure the volume of gas produced.

1

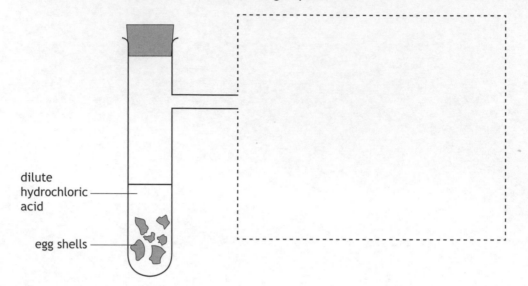

dilute
hydrochloric
acid

egg shells

(b)  Name the salt produced in this reaction.

1

(c)  The volume of gas produced during the reaction was measured.

| Time (min) | Volume of gas (cm³) |
|------------|---------------------|
| 0 | 0 |
| 2 | 47 |
| 4 | 92 |
| 6 | 114 |
| 8 | 118 |
| 10 | 118 |

MARKS

### 3. (c) (continued)

Draw a line graph of volume of gas against time. Use appropriate scales to fill most of the paper.

3

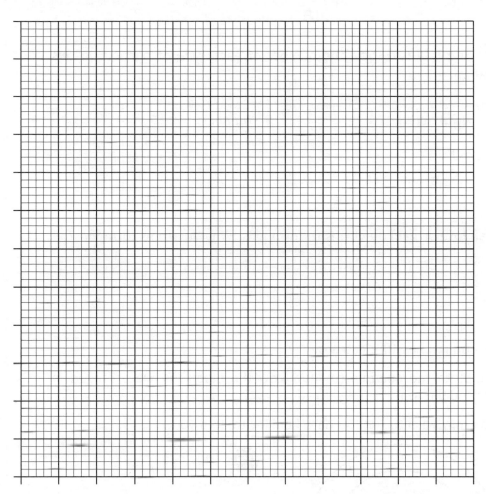

Total marks    5

MARKS

4. Atoms contain particles called protons, neutrons and electrons.

   The nuclide notation of a sodium atom is shown.

   $$^{24}_{11}\text{Na}$$

   (a) Complete the table to show the number of each type of particle in this sodium atom.    1

| Particle | Number |
|----------|--------|
| proton   |        |
| neutron  |        |

   (b) Atoms can lose or gain electrons to form ions. Why do atoms form ions?    1

   (c) An ion of sodium has 10 electrons.

      (i) Complete the diagram to show how the electrons are arranged in this sodium **ion**.    1

      (You may wish to use the data booklet to help you.)

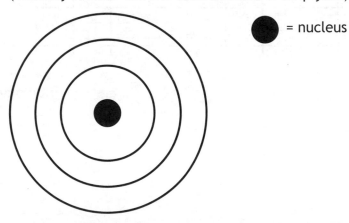

● = nucleus

      (ii) Explain what holds the negatively charged electrons in place around the nucleus.    1

Total marks    4

*Page ten*

MARKS

5. The octane number of petrol is a measure of how efficiently it burns as a fuel.

The higher the octane number, the more efficient the fuel.

(a) The octane numbers for some hydrocarbons are shown.

| Hydrocarbon | Number of carbon atoms | Octane number |
|---|---|---|
| butane | 4 | 90 |
| pentane | 5 | 62 |
| hexane | 6 | |
| heptane | 7 | 0 |
| octane | 8 | −19 |
| 2-methylpentane | 6 | 71 |
| 2-methylhexane | 7 | 44 |
| 2-methylheptane | 8 | 23 |

    (i) Predict the octane number for hexane.    **1**

    (ii) State a relationship between the structure of the hydrocarbon and their efficiency.    **1**

(b) A student investigated the amount of energy released when hexane was burned. The student recorded the following data.

| | |
|---|---|
| Mass of hexane burned | 5 g |
| Volume of water | 1 litre |
| Initial temperature of water | 20 °C |
| Final temperature of water | 78 °C |
| Specific heat capacity of water | 4.18 kJ kg °C$^{-1}$ |

Calculate the energy released, in kJ.    **3**

(You may wish to use the data booklet to help you.)

Total marks    **5**

MARKS | DO NOT WRITE IN THIS MARGIN

6. Read the passage below and answer the questions that follow.

**Ocean Dead Zones**

A dead zone is an area of an ocean (or lake) that has too little oxygen to support marine life; it is hypoxic. This is a natural phenomenon that has been increasing in shallow coastal and estuarine areas as a result of human activities.

Dead zones form due to an increase in nutrients in the water (particularly phosphorus and nitrogen). Human activities have resulted in the near doubling of nitrogen and tripling of phosphorus flows to the environment when compared to natural values.

This dramatic increase in previously limited nutrients results in massive algal blooms. These "red tides" or harmful Algal Blooms can kill fish, cause human illness through shellfish poisoning, and death of marine mammals and shore birds.

This population explosion of algae is unsustainable, and eventually the algae die off, as they block out the light and use up all the oxygen. The algae sink to the bottom, and bacterial decomposition uses the remaining oxygen from the water.

*The passage on the Dead Zone was taken from an article published on "sailorsforthesea.org".*

(a) Suggest which human activities result in nitrogen and phosphorus compounds found in the water.  **1**

(b) Name the term used to describe an area of the ocean that does not have enough oxygen to support life.  **1**

(c) Name the two factors that directly contribute to the low oxygen levels found in these dead zones.  **1**

(d) How does the concentration of phosphorus in 'Dead Zones' compare to natural levels of phosphorus found in water?  **1**

**Total marks    4**

MARKS | DO NOT WRITE IN THIS MARGIN

7. The voltage obtained when different pairs of metal strips are connected in a cell varies and this leads to the electrochemical series.

Using the apparatus below, a student investigated the electrochemical series. Copper and four other metal strips were used in this investigation.

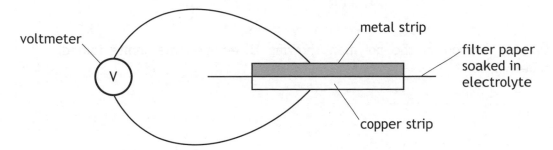

voltmeter

V

metal strip

filter paper soaked in electrolyte

copper strip

The results are shown.

| Metal strip | Voltage (V) | Direction of electron flow |
|---|---|---|
| 1 | 0·6 | metal 1 to copper |
| 2 | 0·2 | copper to metal 2 |
| 3 | 0·9 | metal 3 to copper |
| 4 | 0·1 | copper to metal 4 |

(a) Circle the correct metal to complete the sentence.    1

Connecting metal  1  to metal 2 would produce the largest voltage.
              3
              4

(b) What would be the reading on the voltmeter if both strips of metal were copper?    1

(c) Why can glucose ($C_6H_{12}O_6$) solution **not** be used as the electrolyte?    1

Total marks    3

MARKS

**8.** The monomer in superglue has the following structure.

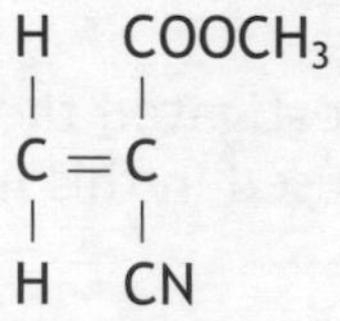

$$\begin{array}{ccc} H & & COOCH_3 \\ | & & | \\ C & = & C \\ | & & | \\ H & & CN \end{array}$$

(a) Draw a section of the polymer, showing **three** monomer units joined together.    1

(b) Name the type of polymerisation that takes place when these monomers combine.    1

(c) Bromine reacts with the monomer to produce a saturated compound.

Draw a structural formula for this compound.    1

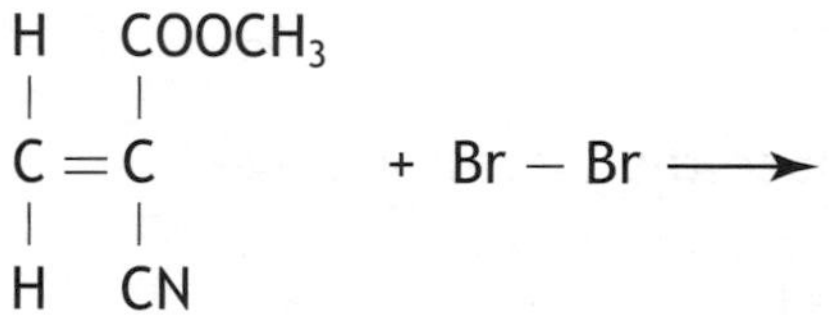

$$\begin{array}{ccc} H & & COOCH_3 \\ | & & | \\ C & = & C \\ | & & | \\ H & & CN \end{array} \quad + \quad Br-Br \longrightarrow$$

Total marks    3

**MARKS** | DO NOT WRITE IN THIS MARGIN

9. Chlorofluorocarbons (CFCs) are a family of compounds that are highly effective as refrigerants and aerosol propellants. However, they are known to damage the ozone layer, and can have a long atmospheric life, which is a measure of how long they exist in the atmosphere.

One example of a CFC molecule ($CCl_2F_2$) is shown.

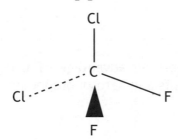

(a) What term is used to describe the **shape** of this molecule? 1

(b) Scientists have developed compounds to replace CFCs. The table shows information about the ratio of atoms in $CCl_2F_2$ and compounds used to replace it.

| Compound | Number of atoms | | | | Atmospheric life (years) |
|---|---|---|---|---|---|
| | C | Cl | F | H | |
| $CCl_2F_2$ | 1 | 2 | 2 | 0 | 102 |
| Replacement 1 | 1 | 1 | 2 | 1 | 13.3 |
| Replacement 2 | 2 | 0 | 4 | 2 | 14.6 |
| Replacement 3 | 1 | 0 | 2 | 3 | 5.6 |

   (i) Draw a structural formula for Replacement 2. 1

MARKS | DO NOT WRITE IN THIS MARGIN

**9. (b) (continued)**

(ii) Compared with $CCl_2F_2$, the replacement compounds contain less of which element? **1**

(iii) From the table, suggest an advantage of using the replacement molecules as refrigerants and aerosol propellants. **1**

**Total marks 4**

MARKS | DO NOT WRITE IN THIS MARGIN

**10.** Silver jewellery slowly tarnishes in air. This is due to the formation of silver(I) sulfide.

The silver (I) sulfide can be converted back to silver using the following apparatus.

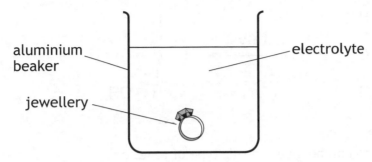

The equation for the reaction that takes place in the beaker is shown.

$$3Ag_2S(aq) + 2Al(s) \longrightarrow 6Ag(s) + Al_2S_3(aq)$$

(a) Calculate the mass, in grams, of silver produced when 0.135 g of aluminium is used up.    **3**

(b) How would you show that aluminium has been lost from the beaker during this reaction?    **1**

**Total marks    4**

MARKS | DO NOT WRITE IN THIS MARGIN

11.    Ethanol is a member of the alcohol family of compounds.

(a)  Ethanol can be manufactured from ethene as shown in the following hydration reaction.

$$
\begin{array}{c}
\text{H}\quad\text{H} \\
|\quad\ | \\
\text{C} = \text{C} \;+\; \text{H}_2\text{O} \;\xrightarrow{\;\text{catalyst}\;}\; \text{H} - \text{C} - \text{C} - \text{H} \\
|\quad\ | \qquad\qquad\qquad\qquad |\quad\ | \\
\text{H}\quad\text{H} \qquad\qquad\qquad\qquad \text{H}\quad\text{OH}
\end{array}
$$

What other name can be given to this type of reaction?                    1

(b)  The following compound is used to give ice cream a rum flavour. It is produced by reacting ethanol with propanoic acid.

$$
\begin{array}{c}
\text{H}\ \ \text{H}\qquad\ \ \text{O}\ \ \text{H}\ \ \text{H} \\
|\ \ \ | \qquad\quad || \ \ \ |\ \ \ | \\
\text{H} - \text{C} - \text{C} - \text{O} - \text{C} - \text{C} - \text{C} - \text{H} \\
|\ \ \ | \qquad\qquad\quad |\ \ \ | \\
\text{H}\ \ \text{H}\qquad\qquad\ \text{H}\ \ \text{H}
\end{array}
$$

(i)   Draw the full structural formula of propanoic acid.                    1

(ii)  To which homologous series does propanoic acid belong?                    1

(c)  Butan-2-ol is another member of the alkanol family.

$$
\begin{array}{c}
\text{H}\ \ \text{H}\ \ \text{H}\ \ \text{H} \\
|\ \ \ |\ \ \ |\ \ \ | \\
\text{H} - \text{C} - \text{C} - \text{C} - \text{C} - \text{H} \\
|\ \ \ \ |\ \ \ |\ \ \ | \\
\text{H}\ \ \text{OH}\ \text{H}\ \ \text{H}
\end{array}
$$

Draw the full structural formula for an isomer of butan-2-ol.                    1

Total marks    4

MARKS | DO NOT WRITE IN THIS MARGIN

**12.** Uranium is a silvery white metallic element that is radioactive because all its isotopes are unstable. Uranium decays by alpha emission.

(a) What is meant by the term isotope?                                                    **1**

(b) Write a balanced nuclear equation for the alpha decay of $^{238}_{92}U$        **2**

(c) Uranium-238 has a half life of 4.5 billion years.

What is meant by the term half life?                                                **1**

**Total Marks    4**

MARKS DO NOT WRITE IN THIS MARGIN

**13.** Read the following passage carefully then answer the question that follows.

### Hydrogen doesn't fit.

The first element, hydrogen, has been causing trouble for some time. It can be placed in group 1, as it usually is, or with the halogens in group 7.

**Periodic Table of the Elements**

| | | | | | | | | | | | | | | | | | | |
|---|---|---|---|---|---|---|---|---|---|---|---|---|---|---|---|---|---|---|
| | | | | | | | | | | | | | | | | | | 2 Helium **He** |
| 3 Lithium **Li** | 4 Berllium **Be** | | Atomic number Name of element Symbol | | | | | | 1 Hydrogen **H** | | | | 5 Boron **B** | 6 Carbon **C** | 7 Nitrogen **N** | 8 Oxygen **O** | 9 Fluorine **F** | 10 Neon **Ne** |
| 11 Sodium **Na** | 12 Magnesium **Mg** | | | | | | | | | | | | 13 Aluminum **Al** | 14 Silicon **Si** | 15 Phosphorus **P** | 16 Sulphur **S** | 17 Chlorine **Cl** | 18 Argon **Ar** |
| 19 Potassium **K** | 20 Calcium **Ca** | 21 Scandium **Sc** | | 22 Titanium **Ti** | 23 Vanadium **V** | 24 Chromium **Cr** | 25 Manganese **Mn** | 26 Iron **Fe** | 27 Cobalt **Co** | 28 Nickel **Ni** | 29 Copper **Cu** | 30 Zinc **Zn** | 31 Gallium **Ga** | 32 Germanium **Ge** | 33 Arsenic **As** | 34 Selenium **Se** | 35 Bromine **Br** | 36 Krypton **Kr** |
| 37 Rubidium **RB** | 38 Strontium **Sr** | 39 Yttrium **Y** | | 40 Zirconium **Zr** | 41 Niobium **Nb** | 42 Molyb-denum **Mu** | 43 Technetium **Tc** | 44 Ruthenium **Ru** | 45 Rhodium **Rh** | 46 Palladium **Pd** | 47 Silver **Ag** | 48 Cadmium **Cd** | 49 Indium **In** | 50 Tin **Sn** | 51 Antimony **Sb** | 52 Tellurium **Te** | 53 Iodine **I** | 54 Xenon **Xe** |
| 55 Caesium **Cs** | 56 Barium **Ba** | 57 Lanthanum **La** | 58–71 ● | 72 Hafnium **Hf** | 73 Tantalum **Ta** | 74 Tungsten **W** | 75 Rhenium **Re** | 76 Osmium **Os** | 77 Iridium **Ir** | 78 Platinum **Pt** | 79 Gold **Au** | 80 Mercury **Hg** | 81 Thallium **Tl** | 82 Lead **Pb** | 83 Bismuth **Bi** | 84 Polonium **Po** | 85 Astatine **At** | 86 Radon **Rn** |
| 87 Francium **Fr** | 88 Radium **Ra** | 89 Actinium **Ac** | 90–103 ■ | 104 Ruther-fordium **Rf** | 105 Dubnium **Db** | 106 Seaborgium **Sg** | 107 Bohrium **Bh** | 108 Hassium **Hs** | 109 Meitnerium **Mt** | | | | | | | | | |

| | | | | | | | | | | | | | | | |
|---|---|---|---|---|---|---|---|---|---|---|---|---|---|---|---|
| ● | 58 Cerium **Ce** | 59 Praseodymium **Pr** | 60 Neodymium **Nd** | 61 Promethium **Pm** | 62 Samarium **Sm** | 63 Europium **Eu** | 64 Gadolinium **Gd** | 65 Terbium **Tb** | 66 Dysprosium **Dy** | 67 Holmium **Ho** | 68 Erbium **Er** | 69 Thulium **Tm** | 70 Ytterbium **Yb** | 71 Lutetium **Lu** |
| ■ | 90 Thorium **Th** | 91 Protactinium **Pa** | 92 Uranium **U** | 93 Neptunium **Np** | 94 Plutonium **Pu** | 95 Americium **Am** | 96 Curium **Cm** | 97 Berkelium **Bk** | 98 Californium **Cf** | 99 Einsteinium **Es** | 100 Fermium **Fm** | 101 Mendelevium **Md** | 102 Nobelium **No** | 103 Lawrencium **Lr** |

Some authors avoid the hydrogen problem altogether by removing it from the main body and by allowing it to float above the rest of the table.

*Passage from RSC.org*

Using your knowledge of chemistry, give reasons why hydrogen can be placed above group 1 or group 7. **3**

MARKS | DO NOT WRITE IN THIS MARGIN

**14.** A student carried out a titration using the chemicals and apparatus shown.

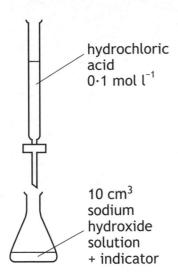

hydrochloric acid
$0.1$ mol $l^{-1}$

10 cm³
sodium
hydroxide
solution
+ indicator

|  | Rough titre | 1st titre | 2nd titre |
|---|---|---|---|
| Initial burette reading/cm³ | 0·3 | 0·2 | 0·5 |
| Final burette reading/cm³ | 26·6 | 25·3 | 25·4 |
| Volume used/cm³ | 26·3 | 25·1 | 24·9 |

(a) Using the results in the table, calculate the **average** volume, in cm³, of hydrochloric acid required to neutralise the sodium hydroxide solution.   1

(b) The equation for the reaction is:

HCl  +  NaOH  →  NaCl  +  H₂O

Using the answer from part (a), calculate the concentration, in mol $l^{-1}$ of the sodium hydroxide solution.   3

**Show your working clearly.**

**Total Marks   4**

**MARKS** | DO NOT WRITE IN THIS MARGIN

**15.** Potassium hydroxide reacts with sulfuric acid to form potassium sulfate, which can be used as a fertiliser.

$$KOH(aq) + H_2SO_4(aq) \rightarrow K_2SO_4(aq) + H_2O(l)$$

(a) Balance the above equation.     **1**

(b) Name the type of chemical reaction taking place.     **1**

(c) Calculate the percentage, by mass, of potassium in potassium sulfate.     **3**

**Show your working clearly.**

**Total Marks     5**

MARKS | DO NOT WRITE IN THIS MARGIN

**ADDITIONAL SPACE FOR ROUGH WORKING AND ANSWERS**

**ADDITIONAL SPACE FOR ROUGH WORKING AND ANSWERS**

MARKS | DO NOT WRITE IN THIS MARGIN

**ADDITIONAL SPACE FOR ANSWERS**

Additional graph paper for Question 3 (c)

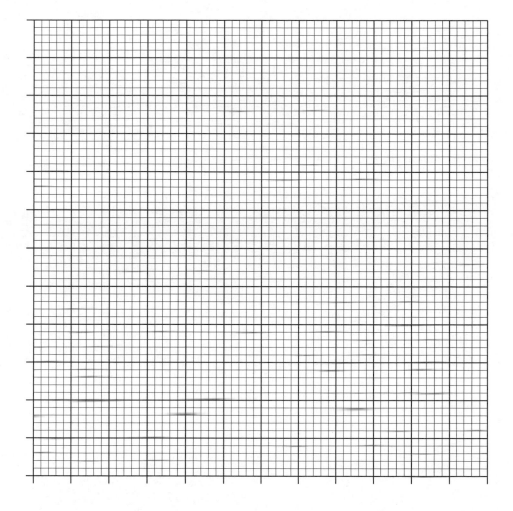

# 2013 Model Paper 2

HODDER
GIBSON
LEARN MORE

# National Qualifications
## MODEL PAPER 2

# Chemistry
## Section 1—Questions

Duration — 2 hours

Instructions for completion of Section 1 are given on Page two of the question paper.

Record your answers on the grid on Page three of your answer booklet.

Do not write in this booklet.

Before leaving the examination room you must give your answer booklet to the Invigilator. If you do not, you may lose all the marks for this paper.

HODDER
GIBSON
LEARN MORE

## SECTION 1

1.  Which of the following elements was discovered before 1775?

    A    Silicon

    B    Oxygen

    C    Bromine

    D    Magnesium

2.  Which of the following is the electron arrangement of sodium metal?

    A    2, 8, 1

    B    2, 8, 2

    C    2, 8, 7

    D    2, 8, 8

3.  Which of the following has a covalent network structure?

    A    Neon

    B    Silicon dioxide

    C    Calcium chloride

    D    Carbon dioxide

4.  The shape of a methane molecule is shown.

    Which of the following compounds would have molecules the same shape as a methane molecule?

    A    Water

    B    Ammonia

    C    Sulfur dioxide

    D    Carbon tetrachloride

5. What is the charge on the copper ion in CuO?

   A    1+

   B    2+

   C    1−

   D    2−

6. Solid ionic compounds do not conduct electricity because

   A    the ions are not free to move

   B    the electrons are not free to move

   C    solid substances never conduct electricity

   D    there are no charged particles in ionic compounds.

7. Which of the following oxides dissolves in water to produce an acidic solution?

   A    $SO_2$

   B    $SiO_2$

   C    $SnO_2$

   D    $PbO_2$

8. Which of the following could be the molecular formula of a cycloalkane?

   A    $C_7H_{10}$

   B    $C_7H_{12}$

   C    $C_7H_{14}$

   D    $C_7H_{16}$

9.

The name of the above compound is

   A    2,2-dimethylbutane

   B    2-ethylpropane

   C    2-methylbutane

   D    3-methylbutane.

**10.** Three members of the cycloalkene homologous series are:

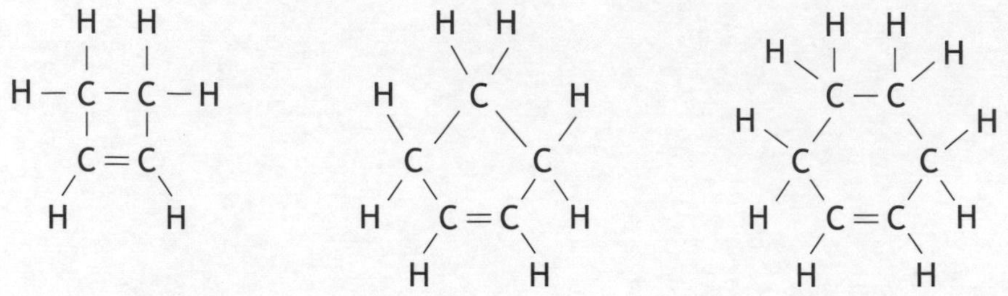

The general formula for this homologous series is

A    $C_nH_{2n+2}$

B    $C_nH_{2n}$

C    $C_nH_{2n-2}$

D    $C_nH_{2n-4}$.

**11.** Which two families of compounds react together to produce esters?

A    Carboxylic acids and alcohols

B    Alkenes and alcohols

C    Alkenes and cycloalkenes

D    Carboxylic acids and cycloalkenes

**12.** What is the correct systematic name for the following compound?

$CH_3CH(CH_3)CH_2CH_2CH_3$

A    Hexane

B    Pentane

C    2-methylpentane

D    3-methylpentane

**13.** Which of the following is an isomer of heptane?

A

```
        H   H   H   H   H
        |   |   |   |   |
    H — C — C — C — C — C — H
        |   |   |   |   |
        H   H   H   |   H
                    |
                H — C — H
                    |
                    H
```

B

```
        H   H   H   H   H   H
        |   |   |   |   |   |
    H — C — C — C — C — C — C — H
        |   |   |   |   |   |
        H   H   H   |   H   H
                    |
                H — C — H
                    |
                    H
```

C

```
                        H
                        |
                    H — C — H
        H   H   H   H   |   H
        |   |   |   |   |   |
    H — C — C — C — C — C = C — H
        |   |   |   |
        H   H   H   H
```

D

```
        H   H   H   H   H   H   H
        |   |   |   |   |   |   |
    H — C — C — C — C — C = C — C — H
        |   |   |   |           |
        H   H   H   H           H
```

14. When propene undergoes an addition reaction with hydrogen bromide, two products are formed.

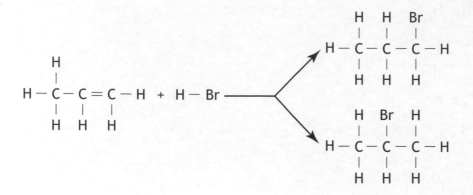

Which of the following alkenes will also produce two products when it undergoes an addition reaction with hydrogen bromide?

A    Ethene

B    But-1-ene

C    But-2-ene

D    Hex-3-ene

15. Experiments were performed on three unknown metal elements, X, Y and Z to establish their reactivity. The results of the experiments are recorded in the table below.

| Metal | Reaction with water | Reaction with dilute acid |
|-------|---------------------|---------------------------|
| X | No reaction | No reaction |
| Y | Slow reaction | Fast reaction |
| Z | No reaction | Slow reaction |

The order of reactivity of the metals, starting with the most reactive, is

A    Y, X, Z

B    X, Z, Y

C    Y, Z, X

D    X, Y, Z.

16. Which of the following diagrams could be used to represent the structure of magnesium?

A

B

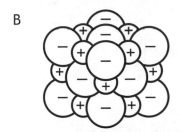

C

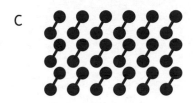

D

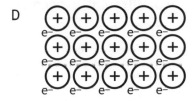

17. Which pair of metals, when connected in a cell, would give the highest voltage and a flow of electrons from **X** to **Y**?

*You may wish to use the data booklet to help you.*

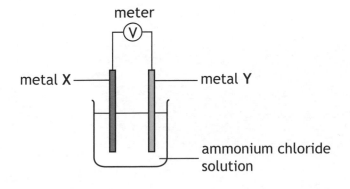

|   | Metal X | Metal Y |
|---|---------|---------|
| A | magnesium | copper |
| B | copper | magnesium |
| C | zinc | tin |
| D | tin | zinc |

18. Which of the following metals would react with zinc chloride solution?

    *You may wish to use the data booklet to help you.*

    A   Copper

    B   Gold

    C   Iron

    D   Magnesium

19. A radioisotope of thorium forms protactinium-231 by beta emission.

    What is the mass number of the radioisotope of thorium?

    A   230

    B   231

    C   232

    D   235

20. Part of the structure of an addition polymer showing two different monomer units combined is shown

$$
\begin{array}{cccccc}
H & H & CH_3 & H & H & H \\
| & | & | & | & | & | \\
-C- & C- & C- & C- & C- & C- \\
| & | & | & | & | & | \\
H & H & H & H & H & H
\end{array}
$$

    Which pair of alkenes could be used as monomers for this polymer?

    A   Ethene and propene

    B   Ethene and butene

    C   Propene and butene

    D   Ethene and pentene

**[END OF SECTION 1.  NOW ATTEMPT THE QUESTIONS IN SECTION 2
OF YOUR QUESTION AND ANSWER BOOKLET.]**

**National Qualifications MODEL PAPER 2**

## Chemistry
## Section 1—Answer
## Grid and Section 2

Duration — 2 hours

**Total marks — 80**

**SECTION 1 — 20 marks**

Attempt ALL questions in this section.

Instructions for completion of Section 1 are given on Page two.

**SECTION 2 — 60 marks**

Attempt ALL questions in this section.

Read all questions carefully before attempting.

**Use blue** or **black** ink. Do NOT use gel pens.

Write your answers in the spaces provided. Additional space for answers and rough work is provided at the end of this booklet. If you use this space, write clearly the number of the question you are attempting. Any rough work must be written in this booklet. You should score through your rough work when you have written your fair copy.

## SECTION 1— 20 marks

The questions for Section 1 are contained in the booklet Chemistry Section 1—Questions. Read these and record your answers on the grid on Page three opposite.

1.   The answer to each question is **either** A, B, C or D.  Decide what your answer is, then fill in the appropriate bubble (see sample question below).

2.   There is **only one correct** answer to each question.

3.   Any rough working should be done on the additional space for rough working and answers sheet.

**Sample Question**

To show that the ink in a ball-pen consists of a mixture of dyes, the method of separation would be:

    A    fractional distillation

    B    chromatography

    C    fractional crystallisation

    D    filtration.

The correct answer is **B**—chromatography.   The answer **B** bubble has been clearly filled in (see below).

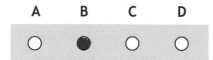

**Changing an answer**

If you decide to change your answer, cancel your first answer by putting a cross through it (see below) and fill in the answer you want. The answer below has been changed to **D**.

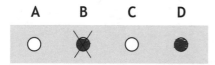

If you then decide to change back to an answer you have already scored out, put a tick (✓) to the **right** of the answer you want, as shown below:

or

## SECTION 1—Answer Grid

|     | A | B | C | D |
|-----|---|---|---|---|
| 1   | ○ | ○ | ○ | ○ |
| 2   | ○ | ○ | ○ | ○ |
| 3   | ○ | ○ | ○ | ○ |
| 4   | ○ | ○ | ○ | ○ |
| 5   | ○ | ○ | ○ | ○ |
| 6   | ○ | ○ | ○ | ○ |
| 7   | ○ | ○ | ○ | ○ |
| 8   | ○ | ○ | ○ | ○ |
| 9   | ○ | ○ | ○ | ○ |
| 10  | ○ | ○ | ○ | ○ |
| 11  | ○ | ○ | ○ | ○ |
| 12  | ○ | ○ | ○ | ○ |
| 13  | ○ | ○ | ○ | ○ |
| 14  | ○ | ○ | ○ | ○ |
| 15  | ○ | ○ | ○ | ○ |
| 16  | ○ | ○ | ○ | ○ |
| 17  | ○ | ○ | ○ | ○ |
| 18  | ○ | ○ | ○ | ○ |
| 19  | ○ | ○ | ○ | ○ |
| 20  | ○ | ○ | ○ | ○ |

[BLANK PAGE]

**MARKS** | DO NOT WRITE IN THIS MARGIN

**SECTION 2— 60 marks**

**Attempt ALL questions.**

1. The Eurofighter "Typhoon" is made from many newly developed materials including titanium alloys.

  (a) The first step in extracting titanium from its ore is to convert it into titanium(IV) chloride.

     Titanium(IV) chloride is a liquid at room temperature and does not conduct electricity.

     What type of bonding, does this suggest, is present in titanium(IV) chloride?     **1**

  (b) Titanium(IV) chloride is then reduced to titanium metal.

     The equation for the reaction taking place is:

     $TiCl_4$   +   $Na \rightarrow$   $Ti$ +   $NaCl$

     (i) Balance the equation.     **1**

     (ii) What does this reaction suggest about the reactivity of titanium compared to that of sodium?     **1**

**Total marks**    **3**

MARKS | DO NOT WRITE IN THIS MARGIN

**2.** (a) Galena is an ore containing lead sulphide, PbS.

　(i) What is the charge on the lead ion in this compound?　　1

　(ii) Calculate the percentage by mass of lead in galena, PbS.　　3

(b) Most metals have to be extracted from their ores.

Place the following metals in the correct space in the table.

　　　copper　　mercury　　aluminium　　　　　　　　　　1

*You may wish to use the data booklet to help you.*

| Metal | Method of extraction |
|---|---|
|  | using heat alone |
|  | electrolysis of molten ore |
|  | heating with carbon |

Total marks　　5

3.  Cool packs can be used to treat some sports injuries.

The pack contains solid ammonium nitrate and water in two separate compartments. When the pack is squeezed the ammonium nitrate dissolves in the water forming a solution. This results in a drop in temperature.

(a)  The change in temperature in the cool pack can be calculated using the equation below.

$$\text{Temperature change} = \frac{\text{energy change (kJ)}}{\text{mass of water (kg)} \times 4 \cdot 2}$$

Calculate the temperature change using the following information.  **2**

| Energy change (kJ) | 6·72 |
| --- | --- |
| Mass of water (kg) | 0·2 |

(b)  Write the ionic formula of ammonium nitrate.  **1**

Total marks  **3**

MARKS | DO NOT WRITE IN THIS MARGIN

4. The element carbon can exist in the form of diamond.

The structure of diamond is shown in the diagram.

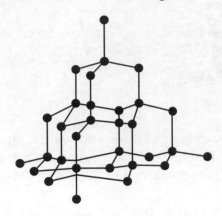

(a) Name the type of **bonding** and **structure** present in diamond.    2

(b) Carbon forms many compounds with other elements such as hydrogen.

Draw a diagram to show how the outer electrons are arranged in a molecule of methane, $CH_4$.    1

Total marks    3

**MARKS** | DO NOT WRITE IN THIS MARGIN

**5.** Volatile organic compounds, VOCs, are organic compounds that can cause damage to the Earth's atmosphere. They may also be harmful or toxic. They are used in paints as solvents and the VOC content is displayed on most paint cans.

(a) An example of a VOC compound used in paints is methanal which is the first member of the aldehydes homologous series. Methanal has the structural formula

What is meant by the term homologous series?    1

(b) Methanal is very flammable. A 2 g sample was burned and the heat produced raised the temperature of 200 cm³ of water from 20.0 °C to 64.7 °C.

Calculate the energy released, in kJ.    3

*You may wish to use the data booklet to help you.*

**Show your working clearly.**

**Total marks    4**

MARKS DO NOT WRITE IN THIS MARGIN

6. A student is given the task of identifying the type of bonding and the elements present in an unknown compound.

Using your knowledge of chemistry, describe tests that the student could perform to identify both the bonding and elements present in the unknown compound. The test descriptions should also include examples of possible results and what the results would indicate. **3**

**Total marks 3**

**MARKS**

**7.** A student was asked to investigate if the type of electrolyte used affects the voltage produced in a cell.

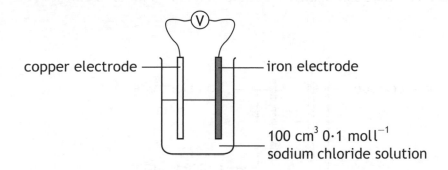

copper electrode — — iron electrode

100 cm$^3$ 0·1 moll$^{-1}$
sodium chloride solution

(a) (i) Complete the labeling of a second cell which could be used to compare the effect of changing the electrolyte from sodium chloride to hydrochloric acid.    **1**

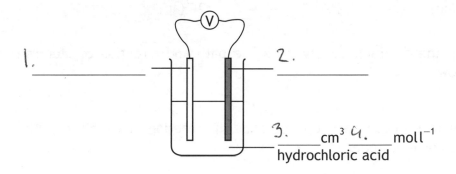

1. _____    2. _____

3. ___ cm$^3$  4. ___ moll$^{-1}$
hydrochloric acid

(ii) Suggest what should be done to make sure the results are reliable.    **1**

MARKS | DO NOT WRITE IN THIS MARGIN

**7. (b) (continued)**

  (b)  Shown is a cell that contains both a metal and a non-metal electrode.

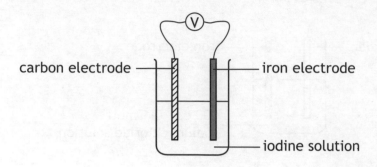

carbon electrode ——|   |—— iron electrode

—— iodine solution

    (i)  The ion-electron equation for the reaction taking place at the carbon electrode is:

$$I_2(aq) \quad + \quad 2e^- \quad \longrightarrow \quad 2I^-(aq)$$

On the diagram, clearly mark the path and direction of electron flow.    1

    (ii)  What term can be used to describe the reaction taking place at the carbon electrode?    1

**Total marks**    **4**

MARKS | DO NOT WRITE IN THIS MARGIN

8. When calcium chloride is dissolved in water, heat is released to the surroundings.

   (a) What term is used to describe chemical reactions which give out heat?    **1**

   (b) A student investigated how changing the mass of calcium chloride affects the heat released.

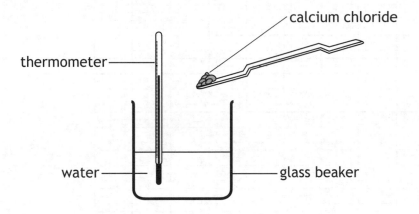

   The results are shown.

| Mass of calcium chloride used (g) | Highest temperature reached (°C) |
|---|---|
| 0 | 20 |
| 5 | 28 |
| 10 | 34 |
| 15 | 41 |
| 20 | 50 |
| 25 | 57 |

MARKS | DO NOT WRITE IN THIS MARGIN

**8. (b) (continued)**

(i)  Plot a line graph of these results.    **3**

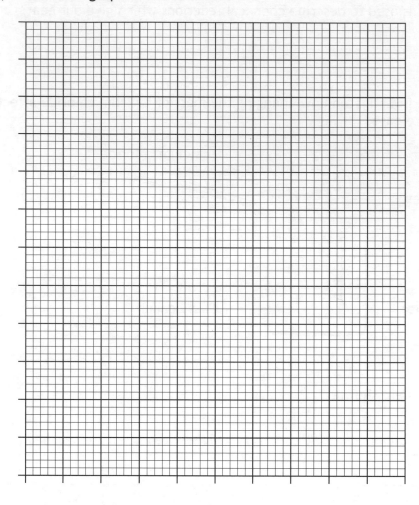

(ii)  Using your graph, find the mass of calcium chloride that would give a temperature of 40 °C.    **1**

(c)  Suggest an improvement that could be made to this experiment.    **1**

**Total marks    6**

MARKS | DO NOT WRITE IN THIS MARGIN

9. Tritium, $_{1}^{3}$H, is an isotope of hydrogen. It is formed in the upper atmosphere when hydrogen atoms capture neutrons from cosmic rays. The tritium atoms then decay by beta-emission.

$$_{1}^{3}\text{H} \quad \rightarrow \quad \qquad + \quad _{-1}^{0}e^{-}$$

(a) Complete the nuclear equation above for the beta-decay of tritium atoms.    **1**

(b) The concentration of tritium atoms in fallen rainwater is found to decrease over time. The age of any product made with water can be estimated by measuring the concentration of tritium atoms.

In a bottle of wine, the concentration of tritium atoms was found to be 25% of the concentration found in rain.

Given that the half-life of tritium is 12.3 years, how old is the wine?    **2**

**Total marks    3**

MARKS | DO NOT WRITE IN THIS MARGIN

**10.** Some sterilising pads contain a 65% solution of Isopropyl alcohol in water.

Isopropyl alcohol has the systematic name propan-2-ol.

(a) Draw the full structural formula of isopropyl alcohol. **1**

(b) Name an isomer of propan-2-ol. **1**

(c) What is the name of the functional group present in all alcohols? **1**

(d) Some sterilising pads also contain ethanol. If a typical pad contains 0·46 g of ethanol, how many moles of ethanol does it contain? **2**

(e) Alcohols can be used to produce esters. What group of compounds must alcohols react with to produce esters? **1**

**Total marks 6**

MARKS | DO NOT WRITE IN THIS MARGIN

11. Read the passage below and answer the questions that follow.

> **Scientists Investigate Release of Bromine in Polar Regions**
>
> Ozone plays a key role not only in the atmosphere, but also on the ground. While at ground level it is not particularly relevant for the protection from UV radiation, it is for the self-cleaning of the atmosphere and removal of contaminants.
>
> In the 1990s researchers discovered that the extensive ozone depletion in the atmosphere close to the ground in the Arctic and Antarctic was due to a reaction of bromine with ozone ($O_3$), producing bromine oxide ($Br_2O$) and oxygen. This bromine is released in autocatalytic processes.
>
> During the polar spring, the resulting bromine oxide clouds can spread over several thousand square kilometers. "It is by far the largest release of bromine on our planet," says Prof. Platt of the Institute of Environmental Physics at Heidelberg University. The precise processes involved are quite complex and are still a topic of current research.

*The passage on Bromine in Polar Regions was taken from an article published on www.uni-heidelberg.de.*

(a) What role does ozone play in our atmosphere?  **1**

(b) Write the formula equation for the reaction of Bromine with ozone. There is no need to balance the equation.  **1**

(c) CFC's such as dichlorofluoromethane are also broken down by UV radiation to produce very reactive free radicals such as chlorine atoms. These chlorine atoms react with the ozone as shown in the equation.

$$Cl(g) + O_3(g) \rightarrow ClO(g) + O_2(g)$$

What mass of ozone would react with 71g of chlorine free radicals?  **3**

**Total marks  5**

MARKS | DO NOT WRITE IN THIS MARGIN

12. The concentration of ethanoic acid in vinegar can be calculated by neutralising a sample with 0.5 mol $l^{-1}$ sodium hydroxide solution.

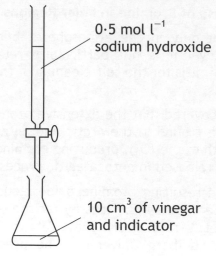

0·5 mol $l^{-1}$ sodium hydroxide

10 $cm^3$ of vinegar and indicator

(a) Draw the full structural formula of ethanoic acid.    1

(b) An average of 20 $cm^3$ of sodium hydroxide was required to neutralise the vinegar.

The equation for the reaction is

$$CH_3COOH + NaOH \rightarrow CH_3COONa + H_2O$$

Calculate the concentration, in mol $l^{-1}$, of the ethanoic acid in the vinegar.    3

(c) Ethanoic acid is classed as a weak acid, but an acid such as hydrochloric acid is classed as a strong acid. This means that a 1·0 mol $l^{-1}$ solution of hydrochloric acid has a pH of 0 but a 1·0 mol $l^{-1}$ solution of ethanoic acid has a pH of 2.

Using your knowledge of chemistry, describe several ways in which the strength of these acids could be compared experimentally.    3

Total marks    7

MARKS | DO NOT WRITE IN THIS MARGIN

**13.** Kevlar is a synthetic fiber that can be used to reinforce the walls of tires.

(a) Kevlar is made from the following monomers.

(i) When these two monomers combine, hydrogen chloride (HCl) is also produced. Draw the structure of the repeating unit formed from these two monomers.

1

(ii) Name the type of polymerisation that takes place to form kevlar.

1

(b) The polymer shown can be used to produce belts for car engines but increasingly car manufacturers are using kevlar as a replacement.

The polymer shown is an example of a polyester. Draw the structural formula of the two monomers that were used to produce this polymer.

2

**Total marks 4**

14. The graph shows how the solubility of potassium chloride changes with temperature.

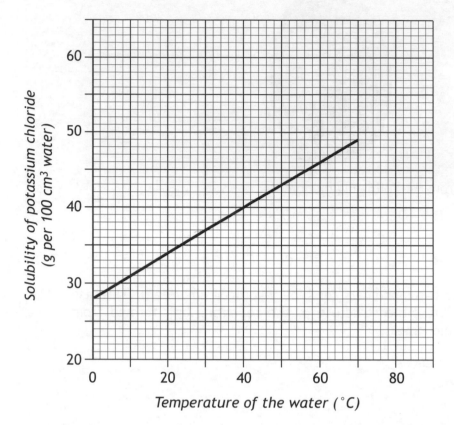

(a) From the graph, what is the maximum mass of potassium chloride that will dissolve at 60 °C?

1

(b) The potassium chloride solution is cooled from 60°C to 30°C. A solid forms at the bottom of the beaker.

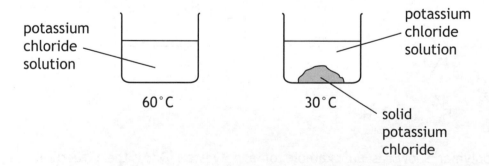

(i) Using the graph, calculate the mass of solid potassium chloride formed at the bottom of the beaker at 30 °C.

2

**MARKS** DO NOT WRITE IN THIS MARGIN

**14. (b) (continued)**

(ii) What method could be used to separate the solid that forms from the potassium chloride solution?

1

**Total Marks    4**

MARKS | DO NOT WRITE IN THIS MARGIN

**ADDITIONAL SPACE FOR ROUGH WORKING AND ANSWERS**

ADDITIONAL SPACE FOR ROUGH WORKING AND ANSWERS

**ADDITIONAL SPACE FOR ANSWERS**

Additional graph paper for Question 8 (b) (i)

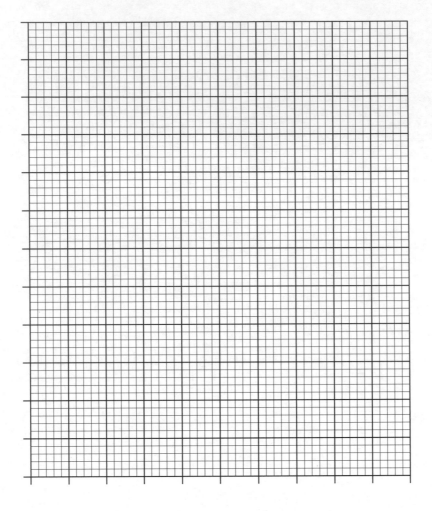

# 2013 Model Paper 3

HODDER
GIBSON
LEARN MORE

# N5

**National Qualifications**
**MODEL PAPER 3**

# Chemistry
## Section 1—Questions

Duration — 2 hours

Instructions for completion of Section 1 are given on Page two of the question paper.

Record your answers on the grid on Page three of your answer booklet.

Do not write in this booklet.

Before leaving the examination room you must give your answer booklet to the Invigilator. If you do not, you may lose all the marks for this paper.

**HODDER GIBSON**
LEARN MORE

## SECTION 1

1. Which of the following has metallic bonding?

   A   Calcium

   B   Carbon

   C   Oxygen

   D   Fluorine

2. The table shows information about an ion.

   | Particle | Number |
   |----------|--------|
   | electron | 19 |
   | neutron | 20 |
   | proton | 18 |

   The charge on the ion is

   A   1+

   B   1-

   C   2+

   D   2−

3. Isotopes of the same element have identical

   A   nuclei

   B   mass number

   C   number of neutrons

   D   number of protons.

4. Which of the following compounds could be used to represent the structure of an ionic compound?

   A

   C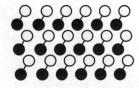

   B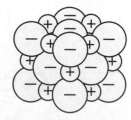

   D

5.  The course of a reaction was followed by measuring the volume of gas produced over time.
    40 $cm^3$ of gas was collected after 100 s.

    What was the average rate of reaction, in $cm^3s^{-1}$, over 100 seconds?

    A    0.04

    B    0.4

    C    2.5

    D    4

6.  The correct formula of calcium sulfite is

    A    $CaS$

    B    $CaSO_3$

    C    $CaSO_4$

    D    $CaS_2O_3$.

7.  Which line in the table correctly shows the properties of an ionic compound?

| | Melting Point (°C) | Conducts Electricity? | |
|---|---|---|---|
| | | Solid | Liquid |
| A | 181 | Yes | Yes |
| B | -95 | No | No |
| C | 686 | No | Yes |
| D | 1700 | No | No |

8.  Compared to pure water, an acidic solution contains

    A    only hydrogen ions

    B    more hydrogen ions than hydroxide ions

    C    more hydroxide ions than hydrogen ions

    D    equal numbers of hydrogen ions and hydroxide ions.

9.  What name is given to the reaction shown by the following equation?

    $$C_2H_4 + H_2 \rightarrow C_2H_6$$

    A    Combustion

    B    Neutralisation

    C    Polymerisation

    D    Addition

10. The balanced equation for the complete combustion of a hydrocarbon X is shown below.

$$X(g) + 2O_2(g) \rightarrow CO_2(g) + 2H_2O(l)$$

Which of the following is the correct formula of hydrocarbon X?

A    $CH_4$

B    $C_2H_6$

C    $C_3H_8$

D    $C_4H_{10}$

11. The table shows the result of heating two compounds with acidified potassium dichromate solution.

| Compound | Acidified potassium dichromate solution |
|---|---|
| 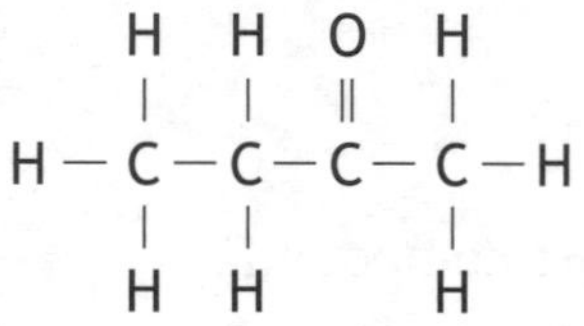 H  H  O  H<br>│  │  ‖  │<br>H — C — C — C — C — H<br>│  │     │<br>H  H     H | stays orange |
| 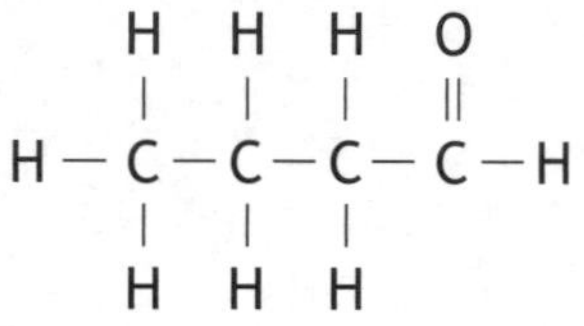 H  H  H  O<br>│  │  │  ‖<br>H — C — C — C — C — H<br>│  │  │<br>H  H  H | turns green |

Which of the following compounds will **not** turn acidified potassium dichromate solution green?

A
```
      H  O  H
      │  ‖  │
  H — C — C — C — H
      │     │
      H     H
```

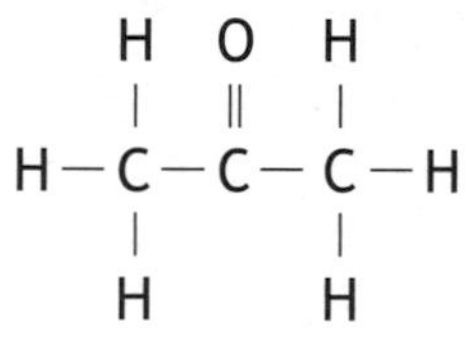

B
```
      H  H  O
      │  │  ‖
  H — C — C — C — H
      │  │
      H  H
```

C
```
      H  O
      │  ‖
  H — C — C — H
      │
      H
```

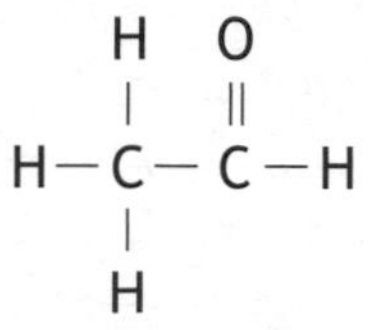

D
```
      O
      ‖
  H — C — H
```

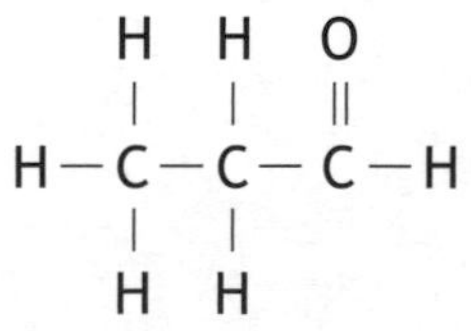

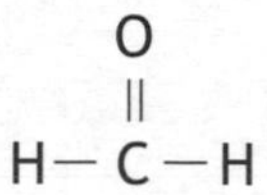

**12.** Which of the following is not the first member of a homologous series?

A

$$H - \underset{\displaystyle \underset{H}{|}}{\overset{\displaystyle \overset{H}{|}}{C}} - H$$

B

(cyclopropane structure with three carbons in a ring, each bonded to two H atoms)

C

$$\underset{\displaystyle \underset{H}{|}}{\overset{\displaystyle \overset{H}{|}}{C}} = \underset{\displaystyle \underset{H}{|}}{\overset{\displaystyle \overset{H}{|}}{C}}$$

D

$$\underset{\displaystyle \underset{H}{|}}{\overset{\displaystyle \overset{CH_3}{|}}{C}} = \underset{\displaystyle \underset{H}{|}}{\overset{\displaystyle \overset{H}{|}}{C}}$$

**13.** Which of the following would quickly decolourise bromine solution?

A    $C_2H_4$

B    $C_3H_8$

C    $C_4H_{10}$

D    $C_5H_{12}$

**14.** $H^+(aq) + NO_3^-(aq) + K^+(aq) + OH^-(aq) \rightarrow K^+(aq) + NO_3^-(aq) + H_2O(l)$

The spectator ions in the reaction are

A    $H^+(aq)$ and $K^+(aq)$

B    $NO_3^-(aq)$ and $OH^-(aq)$

C    $H^+(aq)$ and $OH^-(aq)$

D    $K^+(aq)$ and $NO_3^-(aq)$.

15. The following statements relate to four different metals, **P**, **Q**, **R** and **S**.

    Metal **P** displaces metal **Q** from a solution containing ions of **Q**.

    In a cell, electrons flow from metal **S** to metal **P**.

    Metal **R** is the only metal which can be obtained from its ore by heat alone.

    The order of reactivity of the metals, starting with the **most** reactive is

    A    S, P, Q, R

    B    R, Q, P, S

    C    R, S, Q, P

    D    S, Q, P, R

16. Some metals can be obtained from their metal oxides by heat alone.

    Which of the following oxides would produce a metal when heated?

    A    Calcium oxide

    B    Copper oxide

    C    Zinc oxide

    D    Silver oxide

17. Polythene terephthalate (PET) is used to make plastic bottles, which can easily be recycled by heating and reshaping.

    A section of the PET structure is shown.

    Which line in the table best describes PET?

    |   | Type of polymer | Natural/Synthetic |
    |---|---|---|
    | A | addition | Synthetic |
    | B | condensation | Natural |
    | C | addition | Natural |
    | D | condensation | Synthetic |

**18.** Four cells were made by joining copper, iron, magnesium and zinc to silver.

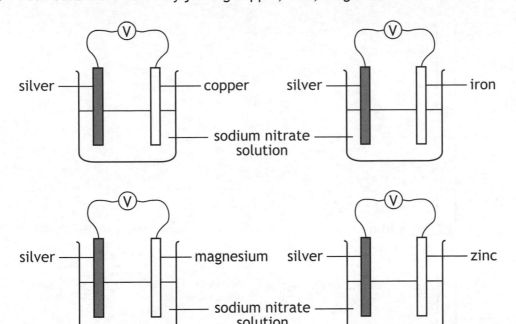

Which of the following will be the voltage of the cell containing silver joined to copper?

*You may wish to use the data booklet to help you.*

A   0.5 V

B   0.9 V

C   1.1 V

D   2.7 V

**19.** Which particle will be formed when an atom of $^{212}_{83}Bi$ emits an alpha particle?

A   $^{207}_{82}Pb$

B   $^{208}_{81}Tl$

C   $^{209}_{80}Hg$

D   $^{210}_{79}Au$

**20.**   In which of the following test tubes will a reaction occur?

A

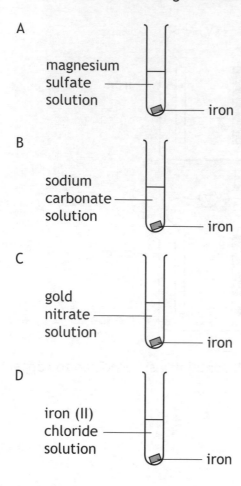

magnesium
sulfate
solution
— iron

B

sodium
carbonate
solution
— iron

C

gold
nitrate
solution
— iron

D

iron (II)
chloride
solution
— iron

**[END OF SECTION 1.  NOW ATTEMPT THE QUESTIONS IN SECTION 2
OF YOUR QUESTION AND ANSWER BOOKLET.]**

## National Qualifications MODEL PAPER 3

# Chemistry
# Section 1—Answer
# Grid and Section 2

Duration — 2 hours

**Total marks — 80**

**SECTION 1 — 20 marks**

Attempt ALL questions in this section.

Instructions for completion of Section 1 are given on Page two.

**SECTION 2 — 60 marks**

Attempt ALL questions in this section.

Read all questions carefully before attempting.

Use **blue** or **black** ink.  Do NOT use gel pens.

Write your answers in the spaces provided.  Additional space for answers and rough work is provided at the end of this booklet.  If you use this space, write clearly the number of the question you are attempting.  Any rough work must be written in this booklet.  You should score through your rough work when you have written your fair copy.

## SECTION 1— 20 marks

The questions for Section 1 are contained in the booklet Chemistry Section 1—Questions. Read these and record your answers on the grid on Page three opposite.

1. The answer to each question is **either** A, B, C or D. Decide what your answer is, then fill in the appropriate bubble (see sample question below).

2. There is **only one correct** answer to each question.

3. Any rough working should be done on the additional space for rough working and answers sheet.

**Sample Question**

To show that the ink in a ball-pen consists of a mixture of dyes, the method of separation would be:

    A    fractional distillation

    B    chromatography

    C    fractional crystallisation

    D    filtration.

The correct answer is **B**—chromatography. The answer **B** bubble has been clearly filled in (see below).

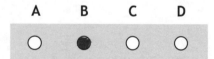

**Changing an answer**

If you decide to change your answer, cancel your first answer by putting a cross through it (see below) and fill in the answer you want. The answer below has been changed to **D**.

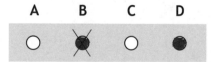

If you then decide to change back to an answer you have already scored out, put a tick (✓) to the **right** of the answer you want, as shown below:

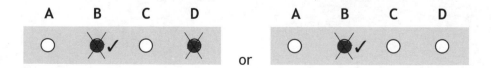

## SECTION 1—Answer Grid

|    | A | B | C | D |
|----|---|---|---|---|
| 1  | ○ | ○ | ○ | ○ |
| 2  | ○ | ○ | ○ | ○ |
| 3  | ○ | ○ | ○ | ○ |
| 4  | ○ | ○ | ○ | ○ |
| 5  | ○ | ○ | ○ | ○ |
| 6  | ○ | ○ | ○ | ○ |
| 7  | ○ | ○ | ○ | ○ |
| 8  | ○ | ○ | ○ | ○ |
| 9  | ○ | ○ | ○ | ○ |
| 10 | ○ | ○ | ○ | ○ |
| 11 | ○ | ○ | ○ | ○ |
| 12 | ○ | ○ | ○ | ○ |
| 13 | ○ | ○ | ○ | ○ |
| 14 | ○ | ○ | ○ | ○ |
| 15 | ○ | ○ | ○ | ○ |
| 16 | ○ | ○ | ○ | ○ |
| 17 | ○ | ○ | ○ | ○ |
| 18 | ○ | ○ | ○ | ○ |
| 19 | ○ | ○ | ○ | ○ |
| 20 | ○ | ○ | ○ | ○ |

[BLANK PAGE]

MARKS | DO NOT WRITE IN THIS MARGIN

**SECTION 2— 60 marks**

**Attempt ALL questions.**

1.  Iron displaces silver from silver(I) nitrate solution.

    The equation for the reaction is:

    $Fe(s) + 2Ag^+(aq) + 2NO_3^-(aq) \rightarrow Fe^{2+}(aq) + 2Ag(s) + 2NO_3^-(aq)$

    (a)  Write the ion-electron equation for the reduction step in the reaction.        1

        You may wish to use the data booklet to help you.

    (b)  This reaction can also be carried out in a cell.

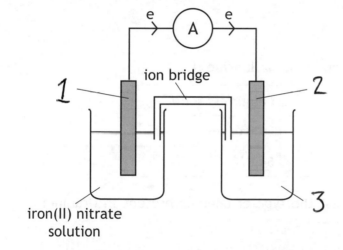

iron(II) nitrate
    solution

        Complete the three labels on the diagram.        1

    (c)  What is the purpose of the ion bridge?        1

                                                        Total marks    3

MARKS

2.  In 2011 33% of Scotland's electricity was produced from nuclear power stations. Coal accounted for 21% with renewable and other sources accounting for the rest.

Green salt (uranium tetrafluoride) can be used to produce fuel for nuclear power stations. It is produced from uranium ore.

(a) Uranium can be extracted from green salt in a redox reaction with magnesium metal.

$$Mg + UF_4 \rightarrow MgF_2 + U$$

Balance the equation.

**1**

(b) This reaction is carried out at temperatures of over 1100°C in an argon atmosphere.

Suggest a reason why the reaction is not carried out in air?

**1**

(c)          **Properties of uranium hexafluoride ($UF_6$)**

| Appearance | Colourless solid |
|---|---|
| Density | 5.09 g/cm$^3$ |
| Melting point | 64.8°C |

Use this information to suggest the type of bonding present in Uranium hexafluoride ($UF_6$).

**1**

(d) When an electric current is passed through water, hydrogen and oxygen are produced. The hydrogen can then be used as a fuel for fuel cells.

Using your knowledge of chemistry, give arguments for and against the suggestion that hydrogen is a pollution free fuel.

**3**

Total marks    **6**

**MARKS** | DO NOT WRITE IN THIS MARGIN

3.  Copper(II) sulfate crystals can be prepared by reacting copper carbonate with dilute sulfuric acid. Water is also produced in the reaction.

(a) Using symbols and formulae, write the chemical equation for this reaction.     1

There is no need to balance the equation.

(b) The four steps involved in the preparation of copper(II) sulfate are shown below:

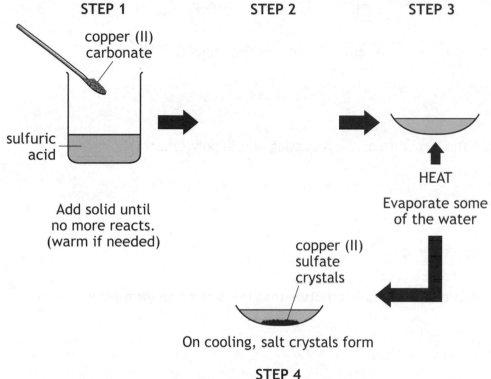

STEP 1     STEP 2     STEP 3

copper (II) carbonate

sulfuric acid

Add solid until no more reacts. (warm if needed)

HEAT

Evaporate some of the water

copper (II) sulfate crystals

On cooling, salt crystals form

STEP 4

Draw the labeled diagram for the 2nd step.     1

(c) In the 'reaction', suggest how you could tell that the reaction is complete.     1

(d) Why must the copper carbonate be added in excess to the sulfuric acid?     1

Total marks     4

MARKS | DO NOT WRITE IN THIS MARGIN

4.   Poly(ethenol) is one of the substances used to cover dishwasher tablets.

A section of the poly(ethenol) polymer is shown.

$$-CH_2-CH-CH_2-CH-CH_2-CH-$$
$$\qquad\;\; |\qquad\qquad |\qquad\qquad |$$
$$\qquad\;\; OH\qquad\quad OH\qquad\quad OH$$

(a)   Name the functional group present in this polymer.     **1**

(b)   Draw the structure of the repeating unit in poly(ethenol).     **1**

(c)   Name the type of polymerisation that takes place to form poly(ethenol).     **1**

Total marks   **3**

MARKS | DO NOT WRITE IN THIS MARGIN

**5.** The diagram shows the apparatus used to prepare chlorine gas.

Concentrated hydrochloric acid is reacted with potassium permanganate.

The gas produced is bubbled through water to remove any unreacted hydrochloric acid and is then dried by bubbling through concentrated sulfuric acid.

(a) Complete the diagram for the preparation of chlorine gas by adding the labels for concentrated sulfuric acid, potassium permanganate and water. **1**

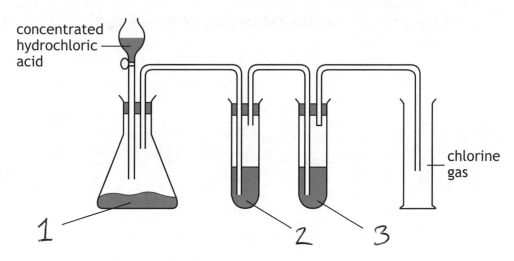

(b) Chlorine is a member of the Group 7 elements.

The graph shows the melting points of these elements.

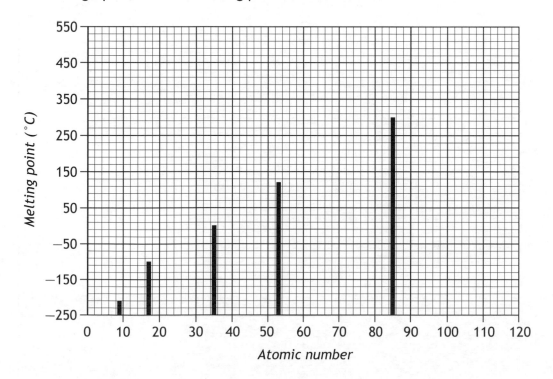

MARKS | DO NOT WRITE IN THIS MARGIN

**5. (b) (continued)**

(i) State the relationship between the atomic number and the melting point of the Group 7 elements.

1

(ii) The next member of this group would have an atomic number of 117.

Using the graph, predict the melting point of this element.

1

**Total marks     3**

MARKS DO NOT WRITE IN THIS MARGIN

**6.** Alkynes are a homologous series of hydrocarbons, which contain carbon to carbon triple bonds. Two members of this series are shown.

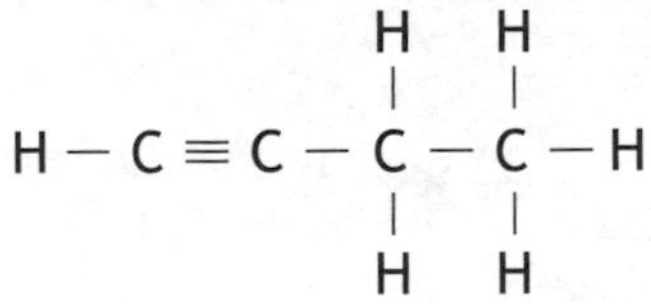

```
        H   H                          H   H   H
        |   |                          |   |   |
H — C ≡ C — C — C — H      H — C ≡ C — C — C — C — H
        |   |                          |   |   |
        H   H                          H   H   H
```

butyne 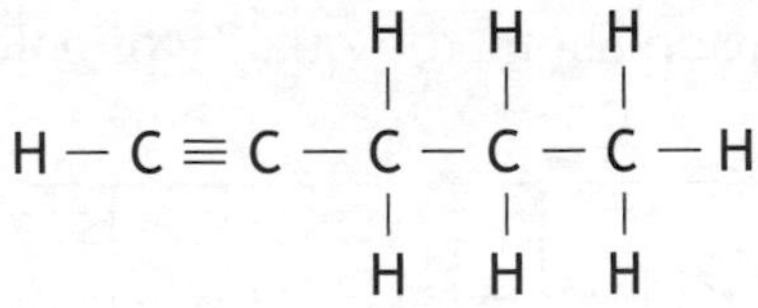                          pentyne

(a) Name the first member of this series.          1

(b) Alkynes can be prepared by reacting a dibromoalkane with potassium hydroxide solution.

```
    H   H   H                          H
    |   |   |                          |
H — C — C — C — H  + 2KOH →  H — C ≡ C — C — H  + 2KBr  + H₂O
    |   |   |                          |
    Br  Br  H                          H
```

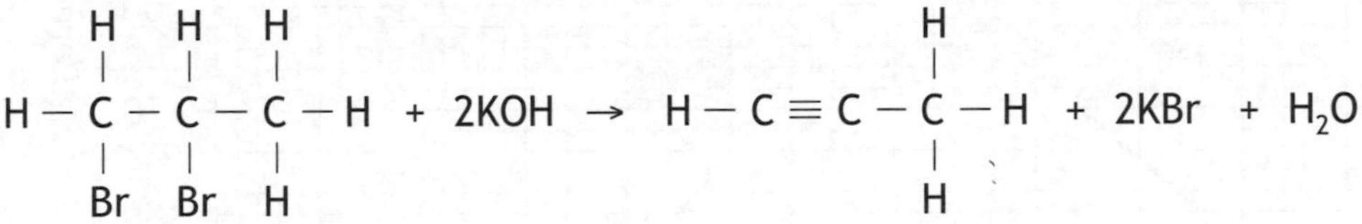

dibromoalkane                          propyne

(i) Draw a structural formula for the alkyne formed when the dibromoalkane shown below reacts with potassium hydroxide solution.          1

```
    H   H   H   H
    |   |   |   |
H — C — C — C — C — H  + 2KOH →
    |   |   |   |
    H   Br  Br  H
```

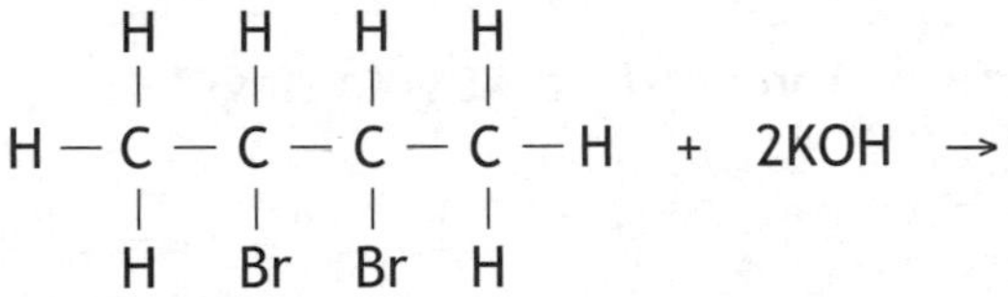

(ii) Suggest a reason why the dibromoalkane shown below does not form an alkyne when it is added to potassium hydroxide solution.          1

```
    H   H   H   H   H
    |   |   |   |   |
H — C — C — C — C — C — H
    |   |   |   |   |
    H   Br  H   Br  H
```

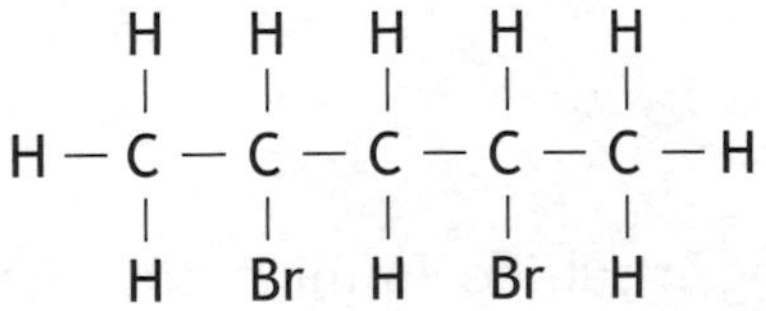

**Total marks   3**

MARKS | DO NOT WRITE IN THIS MARGIN

7. Gold is a very soft metal. In order to make it harder, goldsmiths mix it with silver. The quality of the gold is indicated in carats.

(a) The graph shows information about the quality of gold.

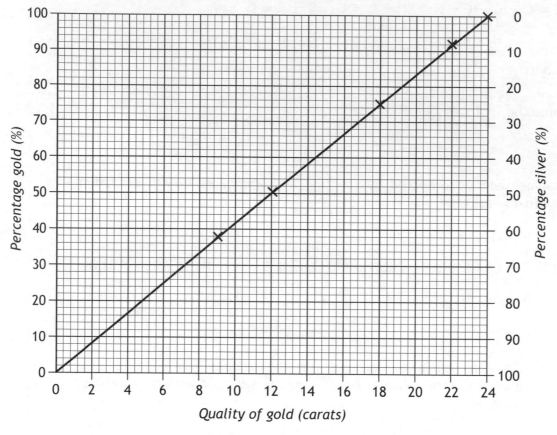

*Quality of gold (carats)*

(i) What is the percentage of silver in an 18 carat gold ring? 1

(ii) Calculate the mass of silver in an 18 carat gold ring weighing 6g. 1

(b) Silver tarnishes in the presence of hydrogen sulfide forming black silver sulfide, $Ag_2S$.

The equation for the reaction is:

$$4Ag + 2H_2S + O_2 \longrightarrow 2Ag_2S + 2H_2O$$

What mass of silver sulfide would be formed from 1·08g of silver? 3

**Total marks** 5

MARKS

8. Chemicals in food provide flavour and smell. Ketones are responsible for the flavour in blue cheese.

Two examples of ketones are shown below.

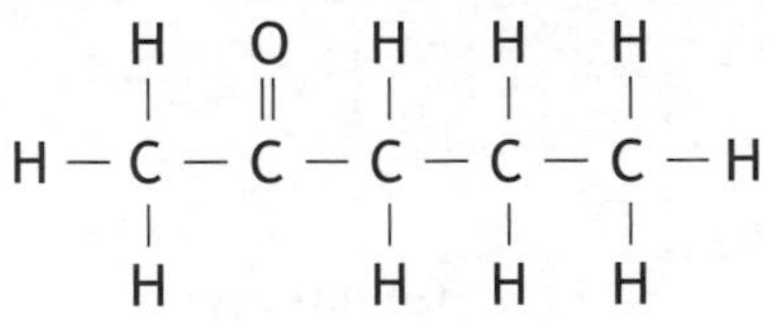

```
   H  O  H  H  H                H  H  O  H  H
   |  ‖  |  |  |                |  |  ‖  |  |
H—C—C—C—C—C—H           H—C—C—C—C—C—H
   |     |  |  |                |  |     |  |
   H     H  H  H                H  H     H  H

   pentan-2-one                 pentan-3-one
```

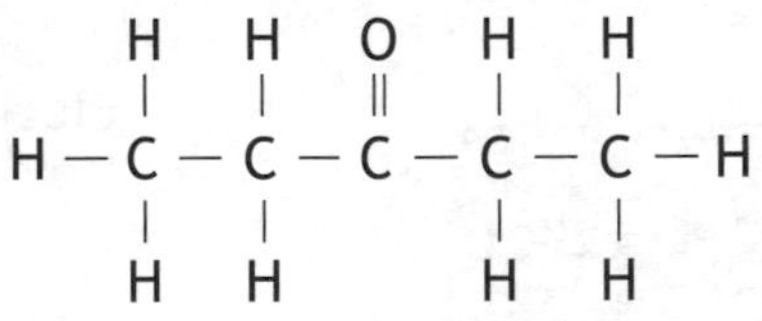

(a) Draw a structural formula for hexan-3-one.    **1**

(b) Suggest a name for the ketone shown.    **1**

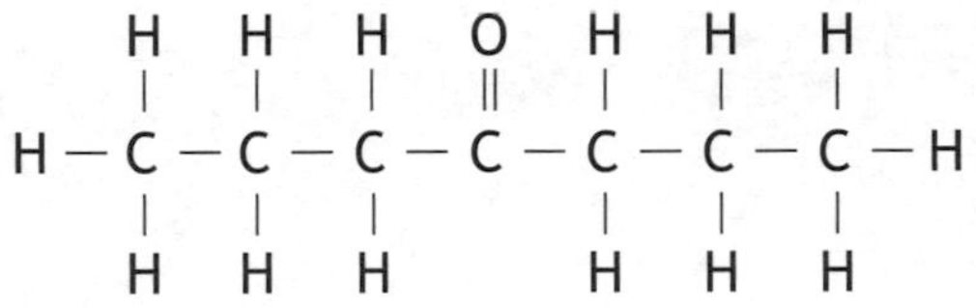

```
   H  H  H  O  H  H  H
   |  |  |  ‖  |  |  |
H—C—C—C—C—C—C—C—H
   |  |  |     |  |  |
   H  H  H     H  H  H
```

(c) Information about the boiling points of four ketones is shown in the table.

| Ketone | Boiling point ($^\circ$C) |
|--------|---------------------------|
| $C_3H_6O$ | 56 |
| $C_4H_8O$ | 80 |
| $C_5H_{10}O$ | 102 |
| $C_6H_{12}O$ | 127 |

Predict the boiling point of $C_7H_{14}O$.    **1**

(d) Sweets, such as pineapple cubes, contain the ester methyl butanoate to provide flavour.

  (i) Give another use of esters.    **1**

  (ii) Methyl butanoate can be broken down to form methanol and butanoic acid.
  Draw the full structural formula of butanoic acid.    **1**

Total marks   **5**

9.  Alpha, beta and gamma radiation is passed from a source through an electric field. The gamma radiation passes directly through, unaffected by the charged plates.

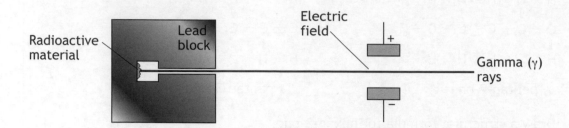

Draw lines on the diagram to show what effect you would expect the charged plates would have on alpha and beta particles. Remember to label each of your lines.

**2**

**Total marks    2**

MARKS

DO NOT WRITE IN THIS MARGIN

10. Read the following passage carefully and answer the questions that follow.

---

**Self Distilling Vodka**

Scientists were investigating the permeability of a material called graphene oxide. This is graphene that has been reacted with a strong oxidising agent, making it more soluble and easier to deal with. They created membranes made up of small pieces of graphene oxide which pile up like bricks to form an interlocked structure, and then tested how gas-proof they were by using the film as a lid for a container full of various gases.

They found that despite being 500 times thinner than a human hair, it completely stopped the gases hydrogen, nitrogen and argon from escaping.

It even stopped helium which, being a tiny single atom will escape from party balloons very quickly, and can even diffuse out through a millimeter of glass. They then tried various liquids, and found similar behaviour for ethanol, hexane, acetone, decane and propanol vapour, but when they tried normal water it behaved as if the membrane wasn't there, escaping at least a hundred thousand times faster than any of the other materials. They think the water is forming a layer one molecule thick between the layers of graphene, blocking the route for everything else, but if it dries out, this gap shrinks and seals up. To make use of this behaviour they put some vodka in the container, and left it for a few days. Normally ethanol evaporates faster than water so vodka gets weaker over time, but with their membrane, which blocked the ethanol, the vodka got stronger and stronger.

---

*Taken from the article "Self Distilling Vodka" by Dave Ansell, published on thenakedscientists.com January 2012.*

(a) Name one gas prevented from escaping by the graphene oxide?　　1

(b) Why is helium found as a single atom?　　1

(c) How much faster was water able to escape compared to the other liquids tested?　　1

(d) "Propanol vapour was also unable to escape through the graphene oxide."

　　Give the correct **systematic name** of the two isomers of propanol that may have been used.　　2

Total marks　5

MARKS | DO NOT WRITE IN THIS MARGIN

11. Glass is made from the chemical silica, $SiO_2$, which is covalently bonded and has a melting point of 1700 °C.

    (a) What does the melting point of silica suggest about its structure?    **1**

    (b) Antimony(III) oxide is added to reduce any bubbles that may appear during the manufacturing process.

    Write the chemical formula for antimony(III) oxide.    **1**

    (c) In the manufacture of glass, other chemicals can be added to alter the properties of the glass. The element boron can be added to glass to make ovenproof dishes.

        (i) Information about an atom of boron is given in the table below.

        | Particle | Number |
        |----------|--------|
        | proton   | 5      |
        | neutron  | 6      |

        Use this information to complete the nuclide notation for this atom of boron.

        # B

        **1**

        (ii) Atoms of boron exist which have the same number of protons but a different number of neutrons.

        What name can be used to describe these different types of boron atoms?    **1**

        Total marks    **4**

MARKS | DO NOT WRITE IN THIS MARGIN

**12.** The flow diagram shows how ammonia is converted to nitric acid.

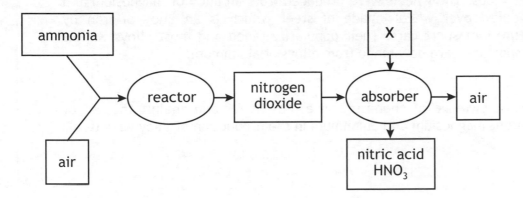

(a) Name the industrial process used to manufacture ammonia.                    1

(b) Name substance **X**.                                                        1

(c) Ammonia and nitric acid react together to form ammonium nitrate, $NH_4NO_3$.
Calculate the percentage by mass of nitrogen in ammonium nitrate.                3
**Show your working.**

Total marks    5

MARKS | DO NOT WRITE IN THIS MARGIN

13.  'Mag Wheels' were a popular type of alloy wheel fitted to sports cars in the 1950s and 1960s. The wheels were produced from an alloy of magnesium and were favoured over wheels made of steel, which is an alloy of iron by manufacturers of sport cars. Their popularity faded and most alloys wheels fitted to sport cars are now made from alloys of aluminium.

Using your knowledge of chemistry, give reasons for and against the use of alloys of iron, magnesium and aluminium in the production of alloy wheels.    **3**

**Total marks    3**

MARKS | DO NOT WRITE IN THIS MARGIN

14. Research is being carried out into making chemicals that can be used to help relieve the side effects of chemotherapy.

One of the reactions in this process is shown

Chemical A + hydrogen    →    Chemical B

(a) As this reaction proceeds, the hydrogen is used up which results in a decrease in pressure.

| Time (min) | 0 | 5 | 10 | 15 | 20 | 30 | 35 | 45 |
|---|---|---|---|---|---|---|---|---|
| Decrease in pressure (bar) | 0 | 0·6 | 1·2 | 1·7 | 2·2 | 2·9 | 3·1 | 3·1 |

Draw a line graph showing the decrease in pressure as time proceeds.    3

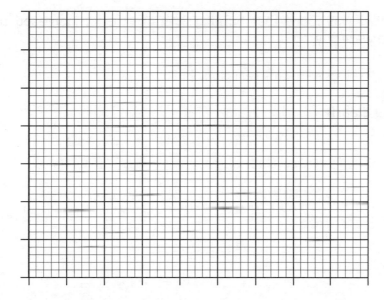

(b) (i) What time did the reaction finish?    1

(ii) Calculate the average rate of the reaction, in bar min⁻¹, between 10 and 20 minutes.    2

Total marks    6

MARKS | DO NOT WRITE IN THIS MARGIN

15. The experiment shown can be carried out to establish how much energy is released when ethanol burns.

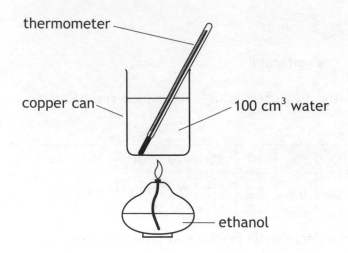

In the experiement it was found that burning 0.1g of ethanol increased the temperature of the water by 7°C.

Calculate the energy released in this reaction, in KJ.    3

**Show your working clearly.**

Total marks    3

**MARKS**

ADDITIONAL SPACE FOR ROUGH WORKING AND ANSWERS

**MARKS** | DO NOT WRITE IN THIS MARGIN

**ADDITIONAL SPACE FOR ROUGH WORKING AND ANSWERS**

**ADDITIONAL SPACE FOR ANSWERS**

Additional graph paper for Question 14 (a)

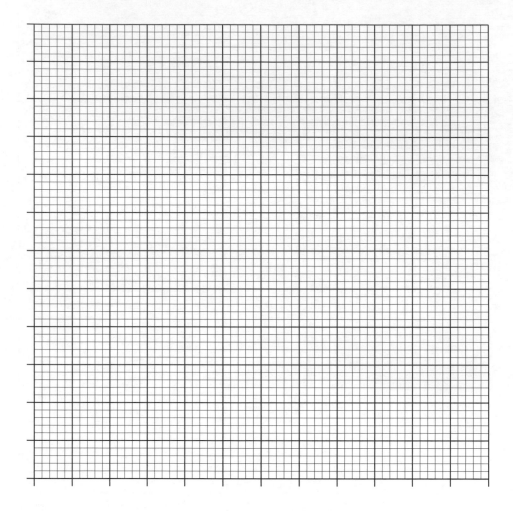

NATIONAL 5 | ANSWER SECTION

# SQA AND HODDER GIBSON NATIONAL 5 CHEMISTRY 2013

## NATIONAL 5 CHEMISTRY SPECIMEN QUESTION PAPER

### Section 1

| Question | Response |
|----------|----------|
| 1 | C |
| 2 | B |
| 3 | A |
| 4 | B |
| 5 | A |
| 6 | D |
| 7 | A |
| 8 | C |
| 9 | C |
| 10 | A |
| 11 | D |
| 12 | C |
| 13 | B |
| 14 | A |
| 15 | C |
| 16 | C |
| 17 | D |
| 18 | D |
| 19 | A |
| 20 | B |

### Section 2

1. (a) 86-88 seconds

   (b) $\dfrac{30 - 0}{20} = 1 \cdot 5$

   (c) Less reactants
   or
   concentration of reactants decreases
   or
   reactants are used up
   or
   less chance of particles colliding
   or
   equivalent answer

2. (a) Atoms with same atomic number/number of protons/positive particles
   and
   different mass number/number of neutrons

   (b) Protons = 35
   Neutrons = 44

   (c) $(79 \times 55) + (81 \times 45) / 100 = 79 \cdot 9$

(d) The maximum available mark would be awarded to a student who has demonstrated a good understanding of the chemistry involved. The student shows a good comprehension of the chemistry of the situation and has provided a logically correct answer to the question posed. This type of response might include a statement of the principles involved, a relationship or an equation, and the application of these to respond to the problem. This does not mean the answer has to be what might be termed an "excellent" answer or a "complete" one.

3. (a) Higher

   (b) (i) Both axes labels with units
   Both scales
   Graph drawn accurately

   (ii) Value must match candidate's graph

4. (a) (i) $Li_2CO_3 + 2HCl \rightarrow 2\ LiCl + CO2 + H_2$

   (ii) Li Cl formula/words/circled /highlighted in equation

   (b) (i) $1/100 = 0 \cdot 01$
   1:1 ratio
   $0 \cdot 01 \times 44 = 0 \cdot 44$

   (ii) Method B

   Gas is lost in method A before starting mass taken
   or
   gas is lost before all acid is added
   or
   no total mass of all reactants at the start of experiment

5. (a) $Al^{3+}(OH^-)_3$

   (b) (i) $16 \times 0 \cdot 7 = 11 \cdot 2$

   (ii) Named active ingredient with appropriate reason. eg
   • magnesium hydroxide − cheapest/doesn't fizz
   • aluminium hydroxide − need to take least amount

6. (a) Any value above 7

   (b) ethene

   (c) Ascorbic acid
   or
   Vitamin C
   or
   benzoic acid

**7.** (a) Group/family/chemicals with same general formula
**and**
same/similar (chemical properties

(b) (i)

$$H : \overset{\displaystyle H}{\underset{\displaystyle H}{\overset{\cdot\cdot}{\underset{\cdot\cdot}{C}}}} : H$$

(ii) Weak bond attraction between molecules
or
Weak intermolecular attractions

**8.** (a) ( O–H )
or
Name of functional group
or
OH written beside question and not circled

(b) (i) addition
or
hydration

(ii) Correct shortened/full structural formula for
3-methylbutan-1-ol
or
3-methylbutan-2-ol

(c) A lot of land used for crops to make ethanol and not
feed people
or
just as harmful to the environment as gasoline
or
low yield
or
deforestation

(d) (i) Correct shortened
or
full structural formula for ethanoic acid
or
Correct mixture of full and shortened formula

(ii) Carboxylic acid/alkanoic acid

**9.** (a) Exothermic

(b) (i) $E_H = cm\Delta T = 4\cdot18 \times 0\cdot2 \times 40 = 33\cdot44$

(ii) Any one from:
heat insulation
repeat to get average
move burner nearer to can
remove tripod and clamp can
stir water
thermometer not touching copper can
use clay triangle on tripod

(c) (i) As the number of carbons increases the energy
released increases
or
As the number of carbons decreases the energy
released decreases
or
The energy $\frac{increases}{decreases}$ as the
number of carbons $\frac{increases}{decreases}$

(ii) Any value from 3520 to 3550

**10.** The maximum available mark would be awarded to a
student who has demonstrated a good understanding
of the chemistry involved. The student shows a good
comprehension of the chemistry of the situation and
has provided a logically correct answer to the question
posed. This type of response might include a statement
of the principles involved, a relationship or an equation,
and the application of these to respond to the problem.
This does not mean the answer has to be what might be
termed an "excellent" answer or a "complete" one.

**11.** (a) gfm = 60
$28/60 \times 100$
$46\cdot6\%$

(b) Speeds up reaction
or
Less energy/temperature/heat required
or
equivalent response

**12.** (a) number of half-lives is 2
¼ of 2 = 0·5 g
0·5 g

(b) short half-life
or
would not last long in the body
or
gamma would go right through body
or
equivalent response

(c) beta
or
$\beta$
or
$_{-1}^{0}e$
or
$_{-1}^{0}\beta$

**13.** (a) gfm 143·5g
$1\cdot435 / 143\cdot5 = 0\cdot01$ mol

(b) $0\cdot5$ mol l$^{-1}$

# NATIONAL 5 CHEMISTRY MODEL PAPER 1

## Section 1

| Question | Response | Mark |
|----------|----------|------|
| 1 | D | 1 |
| 2 | A | 1 |
| 3 | A | 1 |
| 4 | D | 1 |
| 5 | D | 1 |
| 6 | A | 1 |
| 7 | C | 1 |
| 8 | C | 1 |
| 9 | B | 1 |
| 10 | C | 1 |
| 11 | C | 1 |
| 12 | A | 1 |
| 13 | A | 1 |
| 14 | C | 1 |
| 15 | A | 1 |
| 16 | C | 1 |
| 17 | D | 1 |
| 18 | D | 1 |
| 19 | B | 1 |
| 20 | B | 1 |

## Section 2

1. (a) (i) $\dfrac{32 - 12}{8}$

   = 2.5 litres per microsecond

   **2**

   (ii) 4.0 (±1) microseconds

   **1**

   (b) $NaN_3 \rightarrow Na + N_2$

   **1**

   (c) Explosive/
   Highly reactive/very reactive
   or
   Flammable

   **1**

2. The maximum available mark would be awarded to a student who has demonstrated a good understanding of the chemistry involved. The student shows good comprehension of the chemistry of the situation and has provided a logically correct answer to the question posed.

   **3**

3. (a) A workable diagram:
   • Syringe (must have plunger)
   • or Displacement of water into a vertical measuring cylinder/graduated test tube
   • Diagram must not have closed-off tubes

   **1**

   (b) Calcium chloride

   **1**

(c)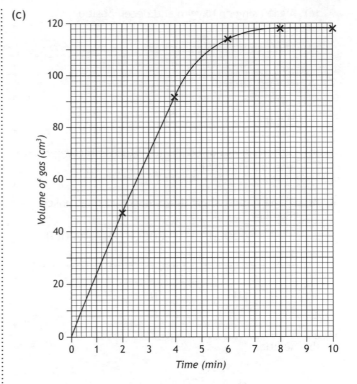

1 mark for correct axes labels and units
1 mark for scale on X and Y axis
1 mark for graph drawn accurately

   **3**

4. (a) 11 − proton
      13 − neutron

   **1**

   (b) To achieve a stable electron arrangement/ full outer energy level/ noble gas arrangement

   **1**

   (c) (i)

   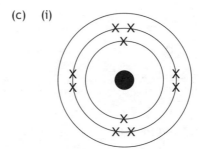

   **1**

   (ii) • The attraction/pull/electrostatic force to the positively charged nucleus (and the (negatively charged) electrons)
   • Attraction/pull/electrostatic force between (positive) protons and electrons

   **1**

5. (a) (i) 25-30

   **1**

   (ii) Branched hydrocarbons burn more efficiently or the greater the number of carbons the lower the octane number efficiency

   **1**

   (b) Eh = cmΔT = 4.18 × 1 × 58 = 242.44 kJ

   **3**

6. (a) Fertilisers

   **1**

   (b) Hypoxic Dead Zone

   **1**

(c) Algae using up oxygen and decomposition

1

(d) Three times as high/higher

1

7. (a) metal 3 circled

1

(b) The reading would be 0 V

1

(c) Glucose does not conduct or glucose is covalent. No ions, only molecules

1

8. (a) Diagram must show three monomer units linked together

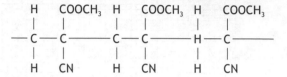

1

(b) Addition

1

(c)

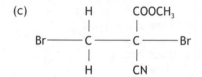

1

9. (a) Tetrahedral/Tetrahedron

(b) (i)

$$F-C-C-H$$ with F, F on left carbon and F, H on right carbon

Or

$$F-C-C-F$$ with H, H on top and F, F on bottom

Or    $CF_3CH_2F$ or $CHF_2CHF_2$

1

(ii) Chlorine/Cl/$Cl_2$

1

(iii) Shorter atmospheric life/breaks down faster

1

10. (a) 1.62g on its own = 3 marks

2 moles Al → 6 moles Ag

54g → 648g

1g → 648 ÷ 54

0.135g → 648 ÷ 54 × 0.135g

= 1.62g

3

(b) Weigh mass of beaker at start and again at the end. (Should have decreased.)
Find mass difference

1

11. (a) Addition / additional

1

(b) (i)
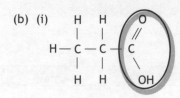

1

(ii) Carboxylic acid

1

(c)
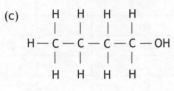
$$H-C-C-C-C-OH$$ with H's on each carbon

Or

$$H-C-C-C-H$$ with H, OH, H on top and H, $CH_3$, H on bottom

Or

$$H-C-C-C-OH$$ with H, H, H on top and H, $CH_3$, H on bottom

1

12. (a) Same atomic number, different mass number

1

(b) $^{238}_{92}U \rightarrow\ ^{234}_{90}Th +\ ^{4}_{2}He$

2

(c) Time take for the activity/ quantity of a radioactive sample to fall by half.

1

13. The maximum available mark would be awarded to a student who has demonstrated a good understanding of the chemistry involved. The student shows good comprehension of the chemistry of the situation and has provided a logically correct answer to the question posed.

3

14. (a) $\dfrac{25\cdot 1 + 24\cdot 9}{2}$

$25\cdot0$ cm³ on its own 1 mark

1

(b) 3 marks 0.25 mol l⁻¹ on its own
1 mole of HCl → 1 mole of NaOH
0.0025 moles of HCl → 0.0025 moles of NaOH
Concentration of NaOH = 0.0025/ 0.01 = 0.25 mol l⁻¹

3

15. (a) $2KOH + H_2SO_4 \rightarrow K_2SO_4 + 2H_2O$

1

(b) Neutralisation

1

(c) 3 marks on its own 44.8%
I mole of $K_2SO_4$ = 174

$\dfrac{78}{174} \times 100 = 44\cdot8\%$

3

## NATIONAL 5 CHEMISTRY MODEL PAPER 2

### Section 1

| Question | Response | Mark |
|----------|----------|------|
| 1 | B | 1 |
| 2 | A | 1 |
| 3 | B | 1 |
| 4 | D | 1 |
| 5 | B | 1 |
| 6 | A | 1 |
| 7 | A | 1 |
| 8 | C | 1 |
| 9 | A | 1 |
| 10 | C | 1 |
| 11 | A | 1 |
| 12 | C | 1 |
| 13 | B | 1 |
| 14 | B | 1 |
| 15 | C | 1 |
| 16 | D | 1 |
| 17 | A | 1 |
| 18 | D | 1 |
| 19 | B | 1 |
| 20 | C | 1 |

### Section 2

**1.** (a) Covalent

1

(b) (i) $TiCl_4 + 4Na \rightarrow Ti + 4NaCl$

1

(ii) Sodium is more reactive than titanium

1

**2.** (a) (i) 2+

1

(ii) 3 marks on its own 86.6%
I mole of PbS = 239

$\dfrac{207}{239} \times 100 = 86.6\%$

3

(b)

| Metal | Method of extraction |
|-------|---------------------|
| mercury | using heat alone |
| aluminium | electrolysis of molten ore |
| copper | heating with carbon |

1

**3.** (a) 2 marks for on its own 8

$\dfrac{6 \cdot 72}{0 \cdot 2 \times 4 \cdot 2} = 8°C$

2

(b) $NH_4^+ NO_3^-$

1

**4.** (a) Covalent bonding/covalent network

2

(b)

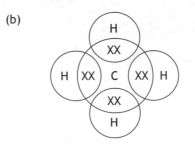

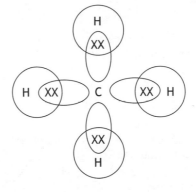

1

**5.** (a) Family of hydrocarbons with similar chemical properties and same general formula

1

(b) 3 marks on its own — 37·4 kJ

Eh = cmΔT = 4·18 × 0·2 × 44·7 = 37·4 kJ

3

**6.** The maximum available mark would be awarded to a student who has demonstrated a good understanding of the chemistry involved. The student shows good comprehension of the chemistry of the situation and has provided a logically correct answer to the question posed.

3

**7.** (a) (i) All three required:
- LHS = copper/Cu
- Top RHS = Iron/Fe
- Bottom RHS = 100 cm³ 0·1 mol l⁻¹

1

(ii) Repeated to allow averages/mean to be calculated

1

(b) (i) From right to left → arrow should be on wires or very close to it

1

(ii) Reduction

1

**8.** (a) Exothermic

1

(b) (i) 1 mark for correct axes labels and units
1 mark for scale on X and Y axis
1 mark for graph drawn accurately

3

(ii) 13 g or correct reading from graph drawn

1

(c) Method of reducing heat loss from beaker to the surroundings / stirring or other suitable answer.

1

9. (a) $^3_1H \rightarrow ^3_2He + ^0_{-1}e$    1

(b) 2 marks for 24·6 years on its own
2 half-lives    2

10. (a)

```
        H   OH  H
        |   |   |
   H — C — C — C — H
        |   |   |
        H   H   H
```
1

(b) Propan-1-ol    1

(c) Hydroxyl    1

(d) 0·01 moles on its own 2 marks
0·46/46 = 0·01    2

(e) Carboxylic acids/ alkanoic acids    1

11. (a) Protection from UV/ removal of contaminants/ self-cleaning of the atmosphere    1

(b) $Br_2 + O_3 \rightarrow Br_2O + O_2$    1

(c) 3 marks 96 g on its own
1 mole of Cl → 1 mole of $O_3$
2 moles of Cl → 2 moles of $O_3$
mass of $O_3$ = 48 x 2 = 96 g    3

12. (a)

```
        H       O
        |      //
   H — C — C
        |      \
        H       OH
```
1

(b) 3 marks 1 mol l⁻¹ on its own
1 mole of NaOH→ 1 mole of $CH_3COOH$
0·01 moles of NaOH→ 0·01 moles of $CH_3COOH$
Concentration of $CH_3COOH$ = 0·01/ 0·01 = 1 mol l⁻¹    3

(c) The maximum available mark would be awarded to a student who has demonstrated a good understanding of the chemistry involved. The student shows good comprehension of the chemistry of the situation and has provided a logically correct answer to the question posed.    3

13. (a) (i)

```
   H         H   O         O
   |         |   ||        ||
 — N —⬡— N — C —⬡— C —
```
1

(ii) Condensation    1

(b)

```
        H  H
        |  |                  O           O
   H—O—C—C—O—H       \\         //
        |  |            C —⬡— C
        H  H          HO           OH
```
2

14. (a) 46 g    1

(b) (i) 46 − 37 = 9 g    2

(ii) Filtration/filter/filtering    1

# NATIONAL 5 CHEMISTRY MODEL PAPER 3

## Section 1

| Question | Response | Mark |
|---|---|---|
| 1 | A | 1 |
| 2 | B | 1 |
| 3 | D | 1 |
| 4 | B | 1 |
| 5 | B | 1 |
| 6 | B | 1 |
| 7 | C | 1 |
| 8 | B | 1 |
| 9 | D | 1 |
| 10 | A | 1 |
| 11 | A | 1 |
| 12 | D | 1 |
| 13 | A | 1 |
| 14 | D | 1 |
| 15 | A | 1 |
| 16 | D | 1 |
| 17 | D | 1 |
| 18 | A | 1 |
| 19 | B | 1 |
| 20 | C | 1 |

## Section 2

1. (a) $Ag^+(aq) + e^- \rightarrow Ag(s)$    1

(b)

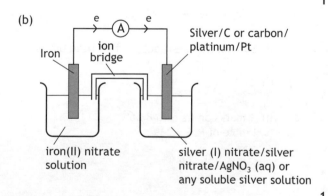

Silver/C or carbon/ platinum/Pt

iron(II) nitrate solution

silver (I) nitrate/silver nitrate/$AgNO_3$ (aq) or any soluble silver solution    1    1

(c) To complete the circuit/allow ions to flow    1

2. (a) $2Mg + UF_4 \rightarrow 2MgF_2 + U$    1

(b) Magnesium would burn/react with oxygen    1

(c) Covalent    1

(d) The maximum available mark would be awarded to a student who has demonstrated a good understanding of the chemistry involved. The student shows good comprehension of the chemistry of the situation and has provided a logically correct answer to the question posed.

3

**3.** (a) $CuCO_3 + H_2SO_4 \rightarrow CuSO_4 + CO_2 + H_2O$

1

(b)

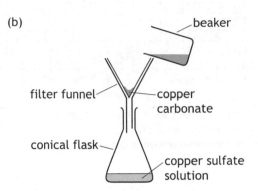

beaker

filter funnel — copper carbonate

conical flask — copper sulfate solution

1

(c) pH neutral, no more gas produced/no more fizzing

1

(d) To ensure that all the acid has reacted.

1

**4.** (a) Hydroxyl

1

(b)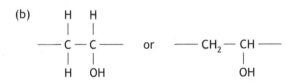

1

(c) Addition

1

**5.** (a) Potassium permanganate
Water
(Conc.) Sulfuric acid

1

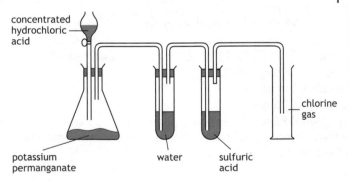

concentrated hydrochloric acid

chlorine gas

potassium permanganate          water          sulfuric acid

(b) (i) As the atomic number increases the melting point increases

1

(ii) 470°C ± 20

1

**6.** (a) Ethyne

1

(b) (i)

$$H-\overset{\displaystyle H}{\underset{\displaystyle H}{C}}-C\equiv C-\overset{\displaystyle H}{\underset{\displaystyle H}{C}}-H$$

1

(ii) Bromines are not attached to adjacent carbon atoms

1

**7.** (a) (i) 25(%)

1

(ii) $25/100 \times 6 = 1\cdot5$ g

1

(b) $1\cdot24$ g on its own 3 marks
4 moles to 2 moles
no of moles of Ag       = $1\cdot08/108$
                                   = $0\cdot01$ moles
no of moles of $Ag_2S$ = $0\cdot01/2$
                                   = $0\cdot005$
GFM $Ag_2S$               = 248
Mass of $Ag_2S$          = $0\cdot005 \times 248$
                                   = $1\cdot24$ g

3

**8.** (a)

$$H-\overset{H}{\underset{H}{C}}-\overset{H}{\underset{H}{C}}-\overset{O}{\overset{\|}{C}}-\overset{H}{\underset{H}{C}}-\overset{H}{\underset{H}{C}}-\overset{H}{\underset{H}{C}}-H$$

1

(b) Heptan-4-one

1

(c) 140–160

1

(d) (i) Solvents or perfumes or materials

1

(ii)

$$H-\overset{H}{\underset{H}{C}}-\overset{H}{\underset{H}{C}}-\overset{H}{\underset{H}{C}}-C\overset{\displaystyle O}{\underset{\displaystyle OH}{}}$$

1

**9.**

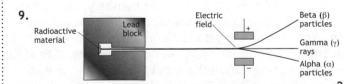

Radioactive material      Lead block      Electric field      Beta (β) particles
Gamma (γ) rays
Alpha (α) particles

2

**10.** (a) Hydrogen, nitrogen, argon and helium

1

(b) Doesn't form bonds/stable electron arrangement

1

(c) Hundred thousand times faster

1

(d) propan-1-ol and propan-2-ol

2

11. (a) Network or Lattice

1

   (b) $Sb_2O_3$

      $(Sb^{3+})_2(O^{2-})_3$

1

   (c) (i) $^{11}_{5}B$

1

      (ii) Isotopes

1

12. (a) Haber

1

   (b) Water/$H_2O$

1

   (c) 35% on its own 3 marks

      $\% = \dfrac{28 \times 100}{80} = 35$

3

13. The maximum available mark would be awarded to a student who has demonstrated a good understanding of the chemistry involved. The student shows good comprehension of the chemistry of the situation and has provided a logically correct answer to the question posed.

3

14. (a)

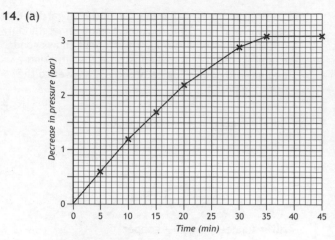

1 mark for correct axes labels and units
1 mark for scale on X and Y axis
1 mark for graph drawn accurately

3

   (b) (i) 35 seconds

1

      (ii) $\dfrac{2 \cdot 2 - 1 \cdot 2}{10}$

         $= 0.1$ bar min$^{-1}$

2

15. Eh = cmΔT = $4 \cdot 18 \times 0 \cdot 1 \times 7 = 2 \cdot 93$ kJ

3

# Acknowledgements

Permission has been sought from all relevant copyright holders and Hodder Gibson is grateful for the use of the following:

An extract taken from an article by Simon Cotton on 'Soundbite molecules' in 'Education in Chemistry' November 2009 © Royal Society of Chemistry (SQP Section 2 page 14);

The passage 'Ocean Dead Zones' taken from an article by Jessica Wurzbacher published on http://sailorsforthesea.org (Model Paper 1 Section 2 page 12);

An extract from the article 'Trouble in the periodic table', taken from page 15 of 'Education in Chemistry', January 2012 © Royal Society of Chemistry (Model Paper 1 Section 2 page 20);

Volatile Organic Compounds Label © B&Q (Model Paper 2 Section 2 page 9);

An extract from the article 'Bromine in Polar Regions' by Dr. Denis Pöhler, Institute of Environmental Physics, University of Heidelberg (Model Paper 2 Section 2 page 17);

An extract taken from the article 'Self Distilling Vodka' by Dave Ansell, published on http://thenakedscientists.com, January 2012 © The Naked Scientists (Model Paper 3 Section 2 page 15).

Hodder Gibson would like to thank SQA for use of any past exam questions that may have been used in model papers, whether amended or in original form.